Compass and Gyroscope

Kai N. Lee

Compass and Gyroscope

Integrating Science and Politics
for the Environment

Foreword by Philip Shabecoff

ISLAND PRESS

Washington, D.C. · Covelo, California

About Island Press

Island Press, a nonprofit organization, publishes, markets, and distributes the most advanced thinking on the conservation of our natural resources—books about soil, land, water, forests, wildlife, and hazardous and toxic wastes. These books are practical tools used by public officials, business and industry leaders, natural resource managers, and concerned citizens working to solve both local and global resource problems.

Founded in 1978, Island Press reorganized in 1984 to meet the increasing demand for substantive books on all resource-related issues. Island Press publishes and distributes under its own imprint and offers these services to other nonprofit organizations.

Support for Island Press is provided by The Geraldine R. Dodge Foundation, The Energy Foundation, The Charles Engelhard Foundation, The Ford Foundation, Glen Eagles Foundation, The George Gund Foundation, William and Flora Hewlett Foundation, The James Irvine Foundation, The John D. and Catherine T. MacArthur Foundation, The Andrew W. Mellon Foundation, The Joyce Mertz-Gilmore Foundation, The New-Land Foundation, The Pew Charitable Trusts, The Rockefeller Brothers Fund, The Tides Foundation, and individual donors.

The author is grateful for permission to reprint excerpts from *Saving the Mediterranean* by Peter M. Haas. Published in 1990 by Columbia University Press. Some arguments and factual material drawn in this book originally appeared in the following publications: "Unconventional Power: Energy Efficiency and Environmental Rehabilitation under the Northwest Power Act" by Kai N. Lee. Reproduced, with permission, from the *Annual Review of Energy and the Environment*. Vol. 16, © 1991 by Annual Reviews Inc. "The Mighty Columbia: Experimenting with Sustainability" by Kai N. Lee. Reprinted with permission of the Helen Dwight Reid Education Foundation. Published by Heldref Publications, 1319 18th Street, N.W., Washington, DC 20036-1802. Copyright 1989.

Library of Congress Cataloging-in-Publication Data

Lee, Kai N.
    Compass and gyroscope : integrating science and politics for the environment / Kai N. Lee.
        p.   cm.
    Includes bibliographical references and index.
    ISBN 1-55963-197-X (cloth).
    1. Sustainable development—Columbia River Watershed.   2. Environmental protection—Columbia River Watershed.   3. Biotic   communities—Columbia River Watershed.   4. Environmental policy—Columbia River Watershed.   I. Title.
HC107.A195L43   1993
333.7'09797—dc20   92-38824
                                                                                        CIP

Printed on recycled, acid-free paper

Manufactured in the United States of America

10 9 8 7 6 5 4 3 2 1

Contents

Foreword

The 1992 Earth Summit in Rio de Janeiro could have been a turning point in the increasingly confrontational relationship between human beings and the natural world. At that gathering, political leaders at the highest level agreed that ecological preservation and economic progress are inseparable and that henceforward we must plan and conduct development so that it is sustainable into future generations. Formal agreements were signed committing nations to address global warming and the loss of biological diversity. Agenda 21—an 800-page, detailed blueprint of environmental, economic, and social actions for achieving sustainable development—was accepted by consensus. Acting on the summit's recommendation, the United Nations General Assembly created a new Commission on Sustainable Development to monitor progress toward the goals of the summit.

Little has changed, however, in the aftermath of Rio, and there are few signs of political will to reach the ambitious goals set there. In the United States, a change of administrations brought a more realistic view of the stakes involved to the highest level of government. Still, there have been no major government actions, no major shifts of policy or economic priorities, here or elsewhere, to start implementing the mandate of the Earth Summit. Nor are there likely to be. The unpleasant truth is that the human community still does not know how to go about the business of creating a way of living that will sustain us and the other life-forms with which we share the planet into future generations. We do not even really know yet what we mean by sustainable development. Until we learn what we want and how to achieve it, the Earth Summit will be remembered not as a turning point but as just another exercise in grandiose geopolitics.

Compass and Gyroscope by Kai N. Lee is a work of extraordinary timeliness. Professor Lee lays out clearly and eloquently not a set of answers to our dilemma but a strategy—a road map—for solving the riddle of sustainable development. Given the messianic certitude of policy advocates and the arrogance of many policymakers, what Lee proposes is almost breathtaking in its modesty: We must have the patience and honesty to *learn* what our real goals should be and how to reach them. By learning he means something new—at least new in the formulation of public policy. His term for it is "adaptive management," deliberate long-term experimentation with the economic uses of ecological systems to learn what works and what doesn't. In terms of defining and achieving sustainable development, for example, we would experience social learning on a global scale. Such an approach is imperative, he argues, because the consequences of developing public policy by trial and error have now become far too dangerous. As he points out, "The appetites of economic growth will eventually reach the limits of the natural world to support them." Exponential population growth and uncontrolled consumption in the industrialized countries are carrying us toward those limits at frightening speed. The prescriptions presented by fervent advocates and politically consecrated problem solvers, however, are likely, without testing and verification, to be the wrong ones.

Such broad-scale social experimentation would appear to be a daunting, almost impossible task. Perhaps, but we have powerful tools—what Lee calls "navigational aids"—to help us on our way. One is the "compass" of physical and social science that can point us in the right direction by applying the rigor of analysis, verification, and correction to our public policies. The other is the "gyroscope" of democratic debate, which subjects the answers offered by science to the rough-and-tumble competition of the free market of interests and ideas, a competition that can keep policy from veering off into the shoals of serious error.

Compass and Gyroscope addresses many of the most vexing issues facing the international community as it attempts to cope with the human impact on the natural world: equity among nations, equity over generations, reaching consensus among broadly conflicting interests, providing political solutions to problems created by private behavior. Professor Lee also tackles the bedrock issue of whether we must redefine our ideas of what constitutes progress. If we cannot sustain the current model of industrial growth forever, perhaps we may have to shift our concept of affluence from having much to wanting little, he suggests.

This is a sophisticated and erudite book that brings the disciplines of ecology, political and social science, and education to bear on what are arguably the most important set of problems now confronting the human community. Environmental activists and militants in developing countries who are convinced that they alone already have the answers to these questions are likely to be impatient with the look-before-we-leap approach advocated by Professor Lee. He does not call for the dispatch of emergency squads to deal with our current disasters. Rather, he is proposing we work hard for a long time so that we and our children will not have to deal with calamity in the future.

Philip Shabecoff

Preface

One of the peculiar commonplaces of our time is the realization that civilized life cannot continue in its present form. In 1987, in a study titled *Our Common Future*, the United Nations World Commission on Environment and Development observed that humans, through their technology, activities, and sheer numbers, now have "the power radically to alter planetary systems." Indeed, "major, unintended changes are occurring in the atmosphere, in soils, in waters, among plants and animals, and in the relationships among all of these. The rate of change is outstripping the ability of scientific disciplines and our current capabilities to assess and advise. It is frustrating the attempts of political and economic institutions, which evolved in a different, more fragmented world, to adapt and cope. It deeply worries many people who are seeking ways to place those concerns on the political agenda." That somber conclusion rests upon an analysis of economic and technical factors.

Without intentional control and modification of human behavior, the economist Robert Heilbroner has warned, what lies ahead is "change forced upon us by external events rather than by conscious choice, by catastrophe rather than by calculation." Yet there has been surprisingly little discussion about how we can or should undertake institutional and political changes of unprecedented scale and durability. Changes are argued to be necessary, and left at that—or proclaimed as a moral imperative, and left at that. This book ventures further, to demonstrate how science and politics can, in the appropriate combination, be enlisted in the search for a sustainable material culture, and to describe cases showing how some elements in this search have been organized and tried out.

At the core of my analysis is the idea of adaptive management. I encountered the concept first in the classroom, as I juggled teaching and government service in the first few months of my appointment to represent the state of Washington on the Northwest Power Planning Council. But I came to understand the significance of adaptive management as a policy actor, not as a professor. This book is accordingly not social science but social engineering—or would be, if we knew enough to link design reliably to result.

Yet in trying to shift my focus from the U.S. Pacific Northwest to the planet as a whole, I came to see as well the perils of advocacy. My brief is not to sell *the* solution, but to present the case for a promising approach, one that is undeniably difficult and expensive in important respects. The obstacles to a sustainable society are hard and heavy, and the levers short and frail. Learning how to move those obstacles is the first step.

Kellogg House
Williamstown, Massachusetts
January 1993

Compass and Gyroscope

Prologue

After Columbus

Five centuries ago Columbus came upon a new world, a rich land, lightly populated by peoples powerless to resist European diseases and firepower. Unlike the Europe Columbus left and the Asia he was looking for, America was an open frontier; there were treasures to be plundered, lands to be cleared and planted, savages to be converted or subdued—resources to be won by anyone with the strength to stake and work a claim. The Old World was settled and ordered, a bounded domain of confined possibilities. America challenged colonists with an errand into the wilderness. For a time, the New World made the entire globe seem limitless.

One hundred years ago the historian Frederick Jackson Turner announced that the census of 1890 demonstrated the "closing" of the American frontier. The discussions he provoked resonated in the founding of the Sierra Club by a California naturalist named John Muir, in the creation of the U.S. Forest Service by President Theodore Roosevelt, and in the geologist John Wesley Powell's ill-starred effort to organize a sustainable agriculture in the arid lands of the West.

What these visionary conservationists sensed at the turn of the last century has become common wisdom as we approach the turn of the next. The ancients had it right after all. The world *is* bounded, an implausible warm blue island in a cold black sky. We foul this nest at our own peril, because in the real world one cannot "light out to get away from sivilizin'," as Huckleberry Finn did. It is that message, as superficial as a poster and as deep as industrialism, that brings to a close the age of Columbus.

Humanity is a powerful force, carving forests into fields and towns, diverting or creating rivers to irrigate deserts, changing the thermal balance of the planet as industrial economies disrupt the

3

metabolism of ocean and atmosphere. The future of the earth is entwined with the human race—not only in the sense that the earth is our home, not yet in the sense that we can control the planet, but already in the sense that human actions influence decisively the habitability of the world for ourselves and all other species. Humans have often used their power to destroy—both by design, as in the case of the hellish oil-well fires of Kuwait, and by accident, as in the case of the ruined soil of the Tigris and Euphrates river basins, salted by centuries of irrigation.

Notwithstanding those conventional wellsprings of self-confidence, technology, the market economy, government, and the perseverance of the individual, human powers have furnished no conclusive evidence that there is intelligent life on earth. Technology has brought a better life to much of humankind, and we hope that progress can continue. But a hope is a good deal less than a plan. Market transactions do not lead automatically to sound environmental outcomes. Unsound ones abound—toxic wastes in groundwater, wanton harvest of wildlife and forests, urban sprawl. There is no world government to discipline the errors of the market, and the conduct of individual nation-states leads many to wonder whether government is more problem than solution. No person, however visionary, however powerful, can live and exercise power long enough to steer the world economy from where it is now headed onto a stable long-term course. Intelligent stewardship of the planet is unlikely to be found at the individual or species level.

Indeed, there may be *no* path from the unstable vitality of the present to a sustainable long-term relationship between humankind and the natural world. Surely one of the messages of the twentieth century to posterity will be that our science and technology persistently outran our ability to govern our expanding capacity to change the world and ourselves. In time, a sustainable order is likely to emerge. But it may well be the stability of the Dark Ages: a miserable sustenance won with what skills survive, taunted by the legends of lost empire.

If there is a better path, it must be found or built by human institutions, organized entities that can act beyond the reach of individuals. Institutions embody ideas too detailed, too disciplined, and too rigid to reflect any single person, however powerful; but they can become the powerful reflection of many overlapping lives, almost all of them individually modest. Yet the history of institutions offers scant hope for intelligent long-term, global-scale planning and management of the problems we must confront and deal with. In 1977 Harvey Brooks, a leader in rethinking the effects of science on society, pointed out that long-term environmental problems pose a special challenge to humanity. Even if the *causes* of environmental problems such as the greenhouse effect could be easily understood, their cure would be difficult when—as is often the case—we have become committed in ways both deep and complex to the activities that cause these problems. Brooks warned that the very fact that our most advanced societies are pluralistic in goals and democratic in governance would make the environmental contradictions of industrialism virtually intractable. To turn open, pluralistic societies toward the disciplined achievement of a single goal—even one so basic as environmental survival—would require social resources and endurance that might be beyond our capacities. Brooks urged this necessity of long-term social learning: the development of the capacity to deal with the uncertain threats to well-being that imbue industrial society, and to sustain the necessary cures for enough time to make a difference. This book attempts to address that need.

The errand into the wilderness turns out to be an odyssey. As we have explored the physical frontiers of the Earth through the visionary lenses of science, we have encountered an ancient human puzzle: to shape a good life from the imperfect and limited means at our disposal, and to do so in the peaceable company of our fellow beings. On his eventful voyage, Homer's Odysseus was protected by a patron goddess, Athena, the symbol of cleverness and intelligence. Humankind's odyssey over the last 500 years has been blessed by its own deities, in the form of scientific technology and democratic responsiveness. In 1492 the distinction between alchemy and chemistry was still being drawn. The notion that magic could be made routine lay in the future, beyond the trial of Galileo and the debates over Darwinian theory. In 1492 the notion that governments should be responsible to those they ruled would have seemed nearly as strange as the possibility that such a form of governance would become the universal hope of humankind. I have come to think of science and democracy as compass and gyroscope—navigational aids in the quest for sustainability. Science linked to human purpose is a compass: a way to gauge directions when sailing beyond the maps. Democracy, with its contentious stability, is a gyroscope: a way to maintain our bearing through turbulent seas. Compass and gyroscope do not assure safe passage through rough, uncharted waters, but the prudent voyager uses all instruments available, profiting from their individual virtues.

Like Odysseus, humanity returns to its home island with new skills, new learning, and a different sense of self and purpose. Like Odysseus' Ithaca, our island is being recklessly used. The myth of the frontier—the chimera created by Columbus's discovery, and only now being recognized for what it is—leads people all over the planet to live beyond the means of nature to sustain them. The famine, disease, and strife affecting three-quarters of the world's people make manifest the stresses of living beyond those means. The health, prosperity, and civil order enjoyed by much of the remaining quarter suggest the possibility of an alternative, although it is one obtained in significant part by transforming the physical assets of the poor into the property of the rich.

As the capacity to control the planet becomes a reality, the aboriginal claim of humans to rule the earth entails more than power. It entails intelligence—what Aldo Leopold called a "land ethic." It is one of the things we still need to learn.

Chapter 1

Taking Measures

... I never ceased to ponder [in the early 1950s] ...
the obvious deterioration in the quality both of
American life itself and of the natural environment. ... Allowed to proceed unchecked, they
spelled—it was plain—only failure and disaster. But
what of the conceivable correctives? ... Would they
not involve hardships and sacrifices most unlikely to
be acceptable to any democratic electorate? Would
they not come into the sharpest sort of conflict with
commercial interests? Would their implementation
not require governmental power which, as of the
middle of the twentieth century, simply did not exist,
and which no one as yet—least of all either of the
great political parties—had the faintest intention of
creating?

—George F. Kennan, *Memoirs*

We've been in bad shape ever since Columbus
landed. ... But that's O.K. You can't go back. We
must live in this modern world and do what we can
to keep it livable.

—Billy Frank, Jr., chairman, Northwest Indian Fisheries

Human activity disrupts environmental stability on a planetary scale.
Spectacular instances—the erosion of the ozone layer, the creation
of long-lived toxic and radioactive wastes—receive increasing public
attention. But it is the mundane momentum that is impressive in the
long run. Humans already appropriate 40 percent of net primary
productivity on land; two of every five beams of sunlight captured
by living things are already in the service of our species. Our demand
for energy, mostly from nonrenewable fossil sources, equals 2 tons of
coal per person per year; each human accounts for more than 300

pounds of steel annually. The world population is expected to double over the next century, with most of the increase occurring in the developing countries. In 1989 a United Nations panel on sustainable development estimated that economic output must rise between five and ten times to keep up with minimal aspirations for betterment. What remains unanswered is how we might double our numbers and quintuple our economic activity without impairing the long-term ability of the natural environment to feed, clothe, house, and inspire our species.

Social Learning

Today, humans do not know how to achieve an environmentally sustainable economy. If we are to learn how, we shall need two complementary sorts of education. First, we need to understand far better the relationship between humans and nature. The strategy I discuss in this book is *adaptive management*—treating economic uses of nature as experiments, so that we may learn efficiently from experience. Second, we need to grasp far more wisely the relationships among people. One name for such a learning process is politics; another is conflict. We need institutions that can sustain civilization now and in the future. Building them requires conflict, because the fundamental interests of industrial society are under challenge. But conflict must be limited because unbounded strife will destroy the material foundations of those interests, leaving all in poverty. Bounded conflict is politics.

This combination of adaptive management and political change is *social learning*. Social learning explores the human niche in the natural world as rapidly as knowledge can be gained, on terms that are governable though not always orderly. It expands our awareness of effects across scales of space, time, and function. For example, we pump crude oil from deep within the earth and ship it across oceans; we burn in a minute gasoline that took millennia to form; with petroleum and its end products we foul water, soil, and air, overloading their biological capacity. Human action affects the natural world in ways we do not sense, expect, or control. Learning how to do all three lies at the center of a sustainable economy.

The Compass

Adaptive management is an approach to natural resource policy that embodies a simple imperative: policies are experiments; *learn from them*. In order to live we use the resources of the world, but we do not understand nature well enough to know how to live harmoniously within environmental limits. Adaptive management takes that uncertainty seriously, treating human interventions in natural systems as experimental probes. Its practitioners take special care with information. First, they are explicit about what they expect, so that they can design methods and apparatus to make measurements. Second, they collect and analyze information so that expectations can be compared with actuality. Finally, they transform comparison into learning—they correct errors, improve their imperfect understanding, and change action and plans. Linking science and human purpose, adaptive management serves as a compass for us to use in searching for a sustainable future.

To see how adaptive management differs from the trial and error by which humans now learn, consider what happens when a tract in the rain forest is logged. Cutting and removing trees tests beliefs about soil erosion, what plants will grow in the cleared space, pollution of the streams that drain the land, and other aspects of that ecosystem's response to logging. If those beliefs are correct, lumber or cleared land can be obtained without permanent damage to the ecosystem's ability to support life, and understanding is affirmed. Unforeseen results, however, usually bring only loss, because people are seldom prepared to infer lessons that are both clear and capable of being checked against others' experience. In contrast, adaptive managers make measurements so that action yields knowledge—even when what occurs is different from what was predicted. Properly employed, this experimental approach produces reliable knowledge from experience instead of the slow, random cumulation gleaned from unexamined error. When reliable learning prevails, a wide range of outcomes is valuable, and unexpected results produce understanding as well as surprise.

Adaptive management plans for unanticipated outcomes by collecting information. Usually, the greater the surprise, the more valuable the information gained. But the costs of information often seem too high to those who do not foresee such surprises. Framing an appropriate balance between predictable cost and uncertain value is a principal task of the chapters ahead.

The Gyroscope

The environment is necessarily shared by us all; but as every littered park reminds us, what is shared by many is typically abused. Reconciling control with the diversity and freedom essential to a democratic society is the task of bounded conflict.

The conflicts that already pervade environmental policy are likely to increase as nations encounter the domestic and international stresses of moving toward sustainable resource use. Conflict is necessary to detect error and to force corrections. But unbounded conflict destroys the long-term cooperation that is essential to sustainability. Finding a workable degree of bounded conflict is possible only in societies open enough to have political competition. In the United States, environmental politics draws its energy from citizen groups, grass-roots nongovernmental organizations (NGOs) that have skillfully engaged governments, businesses, and individuals. American NGOs have in turn influenced environmental activism all over the world, while Europe's parliamentary democracies have been a center of innovation for "green" political parties, a complementary means of articulating, organizing, and empowering environmental concerns.

Political competition is a messy process. Winston Churchill called democracy the worst form of government save for all others. Yet the existence of more or less open competition in political systems is, paradoxically, what bounds conflict in them. Political competition can persist only where there are rules, both unwritten and written. Chief among them is a shared commitment to address important issues through continual debate. In tyrannies, losers are not just down but out, excluded from further decision making by winners who need respect no limits. Like a spinning gyroscope, competition is motion that can stabilize.

People seek individual freedom (and the political competition it fosters) as a fundamental and universal human right. If their aspirations are fulfilled everywhere, the capacity of bounded conflict to correct error may turn unexpectedly into the salvation of our species. Like a gyroscope deep within a vessel, pointing a true course *because* it is independent of crosswinds, conflict bounded by legitimate restraint may yet provide a lodestar for all.

At the global level the search for an environmentally sustainable economy must involve many nations that still lack constitutional democracy, at least for now. As a result, negotiation must often involve limited objectives, without aiming to resolve *all* the differences that divide nations. Finding ways to bound conflict in these circumstances is another major task of the chapters ahead.

Both adaptive management and bounded conflict are essential for social learning to occur. Adaptive management—the compass—is an idealistic application of science to policy that can produce reliable knowledge from unavoidable errors. Bounded conflict—the gyroscope—is a pragmatic application of politics that protects the adaptive process by disciplining the discord of unavoided error. Together they can bring about learning over the decades-long times needed to move from the current condition of unsustainability toward a durable social order.

Large Ecosystems

Social learning is most urgently needed in *large* ecosystems: territories with a measure of ecological integrity that are divided among two or more governing jurisdictions. Large ecosystems present some of the most difficult problems of environmental science and policy. They are complex, often badly damaged, riven by deep-rooted rivalries among several jurisdictions, and essential to the well-being of large populations. Nothing larger than an aquarium or smaller than the planet forms a closed biological system, although some units, such as river basins, are more nearly integral than others. Large ecosystems provide opportunities for learning from and about the real world. Their governance presents challenges of science, management, and politics, often entangled in ways that resist simple approaches. But without some degree of simplification there can be no learning and no transfer from one case to others. Rather than pursuing conceptual neatness, we should study how human institutions deal with the interdependence created when human boundaries cut across ecological continuities. In the case of large ecosystems, pragmatism is a prime virtue: to learn what we can, and to recognize its limits.

What makes an ecosystem "large" is not acreage but interdependent use; the large ecosystem is socially constructed. Rivers nurture fish and plants, water fields and cities, provide transport for trade and sometimes hydroelectricity for industry. Multiple use of a river or other large ecosystem requires trading off qualities that are hard to compare, controlled by or benefiting different people. Social constructs can be difficult to alter, and the boundaries between competing claimants to a natural resource have often produced stalemate rather than problem solving. But an adaptive approach can loosen deadlock with surprising outcomes. The social dynamism of learning can undermine socially constructed stalemate. Although this book focuses on ecosystem management, the subversive flexibility of social learning has wider implications for environmental and public policy.

Large ecosystems are arenas of interdependence. Humans recognize interdependence only when it is more inconvenient not to do so. The institutions of property, privilege, and authority that predate the recognition of interdependence are durable and likely to remain so. Yet large ecosystems are also laboratories of institutional invention. Out of the fractious chronicle of ecosystems ruled by divergent human interests comes most of our small fund of ideas for managing the planet, the largest ecosystem, the one least likely to come under a single government. The cases of the Columbia River, examined in detail in Chapter 2, and of others discussed in Chapter 5 provide important lessons about what sort of governance can be built and how it might be operated. Focusing only on the global case affords a sample of one; by looking at large ecosystems we expand the possibilities for learning.

Large ecosystems offer the possibility of observing cumulative and other large-scale effects. The problems of suburbs cannot be easily projected from the difficulties faced by rural communities. Biological populations can thrive even while many of their individual members perish. Many social and ecological problems become apparent only in settings of sufficient size and complexity. Such places are usually highly imperfect laboratories. No large ecosystem is perfectly matched to any other; what works in one place, in one case, may not transfer. Yet the fact that knowledge must be incomplete is not a reason to disparage what understanding we can glean from these laboratories.

A First World Bias

This is a book mainly about the American experience. But five centuries after Columbus's discovery of a new world, it is obvious that environmental quality is a global issue in which the developing countries must play a central role: they are three-fourths of humanity now, and likely to become a larger fraction still; they are the stewards of the biological treasures of the tropics; and their aspirations for a better life are likely to become the principal engine of human impacts upon the environment.

Sustainable development is a more vexed issue in the Third World than in industrialized countries, first because of the insistent need for progress, and second because of the continuing call for equity. The poor nations as a group have lost ground over the past generation as the social and political tensions of decolonization, changes in international commodity markets, and continued population growth have produced upheaval and corruption more often than investment and economic growth. The pressures of international debts and disrupted domestic economies leave little room to consider long-term environmental goals—and bitter impatience with suggestions from the rich to forgo economic gain.

There are three reasons why I do not put the role of the developing world at the center of analysis here. The first reason is conceptual. In the search for a future in which living standards characteristic of industrial economies can be maintained, it is essential to consider the institutional limitations and possibilities of advanced industrial societies. The second reason is practical. Because industrial economies have made the biggest mistakes and have had the wealth to learn from some of them, their experience is most of what we have to go on. The third reason is speculative. Nations that have changed the economic role and status of women and the life-chances of the old through social insurance have tempered their rate of population growth. We may have reason to expect that sustainable development can also benefit from conditions in nations able to experiment with different modes of living. That is a hopeful reason to learn.

"Sustainable Development": An Elusive Idea

International organizations such as the United Nations now predictably call for "development that meets the needs of the present without compromising the ability of future generations to meet their own needs." Like national "security" or "affirmative" action, sustainable development seems to have become a slogan to elude hard questions: how might the world achieve sustainability, and how would we recognize that we had done so? I offer a partial answer here: social learning enables us to search for sustainability. The crucial test will be learning over time scales of biological significance—learning that lasts long enough to sense the long-run response of the ecosystems we use. Such long-term learning is the acid test of whether a policy is in fact headed toward sustainability. But a process is not a result, nor is the existence of a process the same as the will to use it.

The experience of America also suggests that sustainability cannot be defined simply, because all societies have become economically interdependent. Until the discovery of the "new" worlds of America, Africa, and Australia, commercial peoples such as the merchants of the Silk Road shared the planet—but did not interact—with autarchic societies of hunters and gatherers and those engaged in subsistence agriculture. Such societies, in which trade was a minor factor of material life, often lived in long-run equilibrium with their natural surroundings. The Native Americans of New England, for instance, moved with the seasons, to take advantage of the shifting availability of food over the course of the year. There was little opportunity or reason for families and individuals to accumulate material possessions beyond those they could carry and use. Although the early English colonists viewed them as poor, the Indians did not think of themselves that way: their world was one in which people were able to obtain what they needed. In such a world, sustainability is readily defined because it is not problematic: people live from the land, permanently.

Commerce disrupted that equilibrium. Trade brought links to a vast outer world, the market in which one's products could be sold. Markets over long distances needed something like money: a high-value, compact means of conveying wealth, so that transactions could overcome the inherent awkwardness and inefficiency of barter. Yet once these economic linkages to the outer world were established, it became possible, profitable, and compelling to exploit the land. For the high-value goods that make trade possible also make possible the acquisition of wealth and power.

Sustainability thus became complex and problematic, because local actions now had global consequences, and vice versa. Furs taken by Indians fetched prices at colonial trading posts, and trees could be turned into lumber bound for Europe. Within a century of contact with European settlers, Native Americans were participants in commerce; and by 1700 beaver pelts had become scarce in New England.[1]

This historical sequence is readily recognizable as the beginning of capitalism, although it was not until the time of the American Revolution that Adam Smith's *The Wealth of Nations* gave the emergent economic system a theoretical description.[2] In the centuries since the European discovery of the New World, markets have developed on a planetary scale. The beef served in American fast-food establishments may have been grown on the stubble of a tropical rain forest cleared for pasture. Japanese skyscrapers are erected by means of plywood forms made from ancient trees in Borneo. Beneath the hum and clatter of the industrial city, one can imagine the hoot of the macaque and the thrash of the alligator.

Can there be sustainable use in lands that are linked by markets to purchasers and suppliers far away? In principle, yes: a biological domain could supply a perpetual stream of economic goods, and the revenues earned by their sale could pay for maintaining the productivity of the biosphere. But in the presence of technological change and consumer tastes that swing with the rhythms of mass marketing, stable equilibria have been rare. Since the tropical nations became self-governing after the Second World War they have experienced conditions so often chaotic, so disrupted by forces ranging from domestic politics to international aid, that in most cases it is implausible even to talk of equilibrium.[3]

These limitations lead me to defer a discussion of sustainable development until the final chapter. It is enough, for now, to bear in mind sustainable misery—famine, forests turned to deserts, hyperinflation. These are visions of futures we want to avoid. If there is a way.

A Map

A sustainable relationship between humans and the rest of nature requires a combination that is simple to state and difficult to achieve: Environmental policy should be idealistic about science and pragmatic about politics. Idealism is necessary because knowledge is limited and rigorous science offers the best-known route to reliable knowledge. Pragmatic environmental politics has been demonstrably successful in advanced industrial nations; pragmatism is necessary in a world of nations that fumble their way toward governing a planet.

The first step is to show that a path toward sustainability exists somewhere. This is the task of Chapter 2 and the story of the Columbia River basin in the American Pacific Northwest. Here, the pursuit of sustainable development has been successfully launched. Moving toward sustainability is a two-part process: first, revising the uses of the ecosystem so that environmental values take an economically relevant place in policy and practice; and second, incorporating the well-being of the ecosystem into the way management is conceived and implemented. In the large ecosystem that is the Columbia River basin, the practical goal is balanced management to produce hydroelectric power while maintaining a habitat suitable for salmon—a balance between the most economically important resource of the river and the most emotionally compelling symbol of its natural integrity.

Chapters 3 and 4 use the lessons of the Columbia to describe the social learning needed to search for sustainability: Adaptive management, discussed in Chapter 3, is a synthesis of science and policy that treats policies as large-scale experiments. Bounded conflict, discussed in Chapter 4, is a combination of politics, negotiation, and other means of promoting uncomfortable change, which provides tools for establishing shared goals and probing the bounds of cooperative effort. Like compass and gyroscope, the two parts of social learning are complementary, each compensating for the weaknesses of the other, the whole standing stronger as a consequence. In Chapter 5 I review the limited body of empirical knowledge about the governance of large ecosystems and other social analogues. Concepts of learning in human institutions and their relevance to large ecosystems are explored in Chapter 6.

Chapter 7 discusses a blend of scientific idealism and political pragmatism that I call "civic science." Civic science, with all its frailties, defines our capacity to move from where we are today toward a sustainable future. Chapter 8 returns to the puzzle of sustainable development and suggests that what is needed is not so much a definition as a perspective on human communities and the world they share.

The following chapters offer no simple answers; they are neither recipe nor cookbook but an essay on cooking. If achieving sustainability were a simple task, social learning and its conflict and confusion would not be necessary. Today, we do not know whether it is possible to achieve sustainability nor how to do it. The environmentalist in me doubts that growth of the kind pursued by rich societies can benefit the planet as a whole. The scientist in me believes that sustainability involves practical puzzles that cannot be ignored by the missionary zeal of environmental advocates or by the technological optimism of those who pursue growth. The tension between scientific truth and the quest for a just society lies at the heart of sustainable development; it is a tension we face through social learning. Conceptual complexity and the tenacity of people make the search for sustainable development difficult. That people have beliefs is not the problem: without beliefs and values, there would be scant reason to help the race endure. Yet the ability of human institutions to learn is frail. We need prudence, inventiveness, and persistence. Fashioning from these elusive, refractory materials a seaworthy vessel for the voyage ahead remains a task for which understanding is insufficient. I claim only that it is necessary.

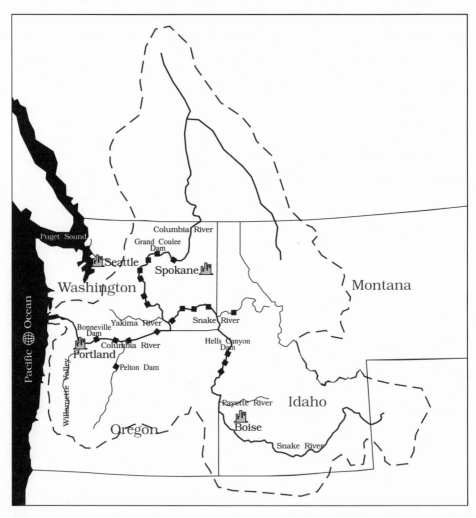

FIGURE 2–1. The Columbia River basin. (Source: Northwest Power Planning Council.)

Chapter 2

Sustainability in the Columbia Basin

To direct attention today to technological affairs is to focus on a concern that is as central now as nation building and constitution making were a century ago.

—Thomas Parke Hughes, *Networks of Power*

When an electric light is turned on in Seattle, a salmon comes flying out. When a room is air-conditioned in Los Angeles, they're doing that on the back of the salmon.

—Billy Frank, Jr., chairman, Northwest Indian Fisheries

Rising in the Canadian Rockies and flowing 1,200 miles through the Pacific Northwest, the Columbia is the fourth largest river in North America. It drains an area that includes parts of seven U.S. states and two Canadian provinces (see Fig. 2–1). The river's average annual streamflow of 141 million acre-feet is more than ten times that of the Colorado. The Columbia's high flows and extensive drainage have made it ideal for colonization, first by fish and wildlife as the glaciers retreated at the end of the last ice age, and, much later, by dam-building humans.

Two Columbia River Civilizations

There have been two Columbia River civilizations; a third is now emerging. To its earliest human inhabitants the river provided a rich life of fishing and gathering centered on the Pacific salmon. Later the river became an industrial waterway, the sinew and pulse of a regional economy. Now the Columbia has become the locus of an awkward search for balance among its industrial uses and its environmental treasures.

Wilderness

In the Columbia River basin just over a century ago, the salmon were more important than money is today. The river is a major spawning ground and nursery of the Pacific salmon. What humans call the river's basin is the landward redoubt of a salmon empire reaching far into the North Pacific Ocean, where the fish mature for two to four years before returning to their native streams to reproduce. After the glaciers retreated roughly 8,000 years ago, this ecosystem supported a population of perhaps 50,000 Native Americans, whose world centered on the yearly migrations that brought 10 to 16 million salmon back to the river. Harvested by spear, net, and boat, these fish provided both food and trade goods for the people of the river basin. The tribes lived in a long-run ecological equilibrium, fluctuating between bad times and good but enduring over many human generations.

That river and the fish it nurtured linger in the memory of Harold Culpus of the Warm Springs Nation of central Oregon:

> Sunday was the time that he worshipped—sometimes Friday, Saturday, Sunday—and they had to have that salmon as part of the traditional way, as part of the ceremony.
> In the beginning, from the time of creation of the Indian as it is told . . . it is interpreted that food is the most sacred thing in this country—food, love and relating a person, brother or sister. . . .
> That's why the Indian, when the white man first came to this country, in the Columbia River, he spread the table out and set the food out for him, though he was a stranger.

This original Columbia civilization lasted until the early nineteenth century, when the strangers arrived.

They came in search of Manifest Destiny, a vision of a nation stretching from sea to shining sea. With them came a pattern repeated since European expansion began in the fifteenth century. Diseases unknown among the Native Americans—against which their bodies were defenseless—decimated the population. The settlers met little resistance as they moved over earth invisibly scorched by smallpox and other plagues. Yet as logging, mining, and farming transformed the landscape, conflict between the remaining Indians and settlers escalated. The winners recall it this way:

> Remember the trial when the battle was won,
> The wild Indian warriors to the tall timber run.
> We hung every Indian with smoke in his gun.
> Roll on, Columbia, roll on.

In 1855 Isaac Stevens, territorial governor and plenipotentiary of the distant powers of the United States of America, concluded treaties with tribes throughout the Pacific Northwest to secure the newcomers' property claims. These treaties created reservations within which the native peoples agreed to live while retaining rights to fish, hunt, and gather roots and plants over a territory well beyond the reservation boundaries "in common with" the settlers. That language would reverberate more than a century later. At the time it seemed not much to concede: it afforded the Indians their traditional livelihood, and there was plenty to share "in common." Governor Stevens had underestimated the powers of his own people.

Beginning in 1969, after the settlers and their descendants had transformed the landscape and obliterated many of the fish runs, the Northwest tribes filed, and won, lawsuits to claim their treaty rights. The immediate result was to reallocate shares of the salmon harvest: Native Americans were entitled to harvest half the fish. Such a drastic and sudden curtailment of a fishing industry already in decline struck hard at commercial and sport fisheries, which had ignored the Indians since the Stevens treaties were signed. After a decade of hard feelings, as the lawsuits made their way to the Supreme Court—where the treaty claims were affirmed—both non-Indian and tribal leaders realized that there was only one option all could abide: to rebuild the salmon populations so that there would once again be enough for all to take "in common" without battling one another for the right to kill off the stocks forever.

The long-term result of the tribal fishing rights dispute was the discovery that sustainable development was indispensable. A principle of wilderness, to take what one needs but to leave enough for the future, had survived industrialization. It remains to be seen whether the principle can endure as policy.

Industry

In 1941 a young folksinger in Portland, Oregon, wrote a song for the new Bonneville Power Administration, the federal agency that had begun to sell power from the just-completed dams.

> . . . roll along, Columbia, you can ramble to the sea,
> But river, while you're rambling, you can do some work for me.

Woody Guthrie was celebrating the second Columbia basin civilization.

The basin's 19 major dams, together with more than five dozen smaller hydro projects, today constitute the world's largest hydroelectric power system. The Columbia River and its tributaries generate on average about 12,000 megawatts from falling water—more power than is used in New York City.

Largely built by the U.S. government at a time of low labor costs, the dams fostered economic growth in the Pacific Northwest with cheap electricity. Industrial and agricultural development have built the population centers of the Northwest: Portland and the Willamette Valley of Oregon, Boise and Spokane in the upper watershed, as well as Seattle and the Puget Sound. Aluminum for soft-drink cans and Boeing airliners comes from Northwest smelters powered by hydropower. So do McDonald's French fries, processed in the potato country of Idaho; plywood and computer chips from Oregon; grocery bags and software from Washington—products and jobs from the river's electricity.

The Indians who lived in wilderness have given way to a population of 9 million, more than 100 times the aboriginal level. So large an increase in population entails a fundamental change in the relationships between people and the environment. The domesticated river provides power and irrigation while dams control its once-destructive floods; locks turn its rapids into an inland waterway navigable by tug and barge for 500 miles from Astoria, Oregon, near the river's mouth to Lewiston, in central Idaho, even as the steady winds of the Columbia Gorge afford world-class windsurfing.

The industrial Columbia is a multiple-purpose marvel, a river, as the historian Donald Worster put it, that died and was reborn as money.[1] The governing principle behind the many functions of the river has been to maximize economic return. The river's uses have been ranked accordingly: power first; then urban and industrial uses, agriculture, flood control, navigation, recreation; and finally fish and wildlife. As a consequence of these priorities, by the late 1970s the salmon runs of 10 to 16 million in the preindustrial era had dwindled to 2.5 million.

Larry Mills, a man whose pale blue eyes gleam with the optimism of Manifest Destiny, was born and raised in the Payette River country of southern Idaho. He remembers fall in his boyhood as a stinky time, when the salmon returning from the Pacific would dig nests in the river gravels, deposit and fertilize their eggs, and die. For a month the reek of rotting salmon would clot the air. Like the older members of the Yakima Indian Nation, who recall streams so densely packed that one could walk from one shore to the other on the bodies of spawned-out fish, Mills had lived to see the valley of his youth stripped of salmon, their return blocked when the Idaho Power Company built Hells Canyon Dam more than 100 miles downriver. Larry Mills was elected from three different districts in Idaho and twice rose to become speaker of the state house of representatives. In retirement, he turned his wily wisdom to the fish he remembered—and found himself an ally of Indians, who knew Manifest Destiny only as a battle cry of conquest.

The appetites of economic growth will eventually reach the limits of the natural world to support them. In the Pacific Northwest, the limits have been measured by fish and power, each an emblem of its age. Salmon, ranging the length of the river since time immemorial, are an indicator of the health of the ecosystem; their decline has been a slow crisis by human calendars, the biological depletions hidden and deferred by shifting fish from Indians to commercial fleets and canneries, from natural habitat to hatcheries. The problems of hydropower, the symbol and substance of modernity, mounted swiftly. The costs involved were high, and the leading alternative to hydro was nuclear power, an energy source that fueled political conflict more brightly than it did lightbulbs. Once the crises of power and fish converged, the industrial era on the Columbia came to an end, although no one knew it at first, and no one can yet say what has succeeded it.

Dwindling Fish Runs

Salmon in the Columbia declined from many causes: overfishing, careless logging and mining that destroyed spawning grounds, industrial and urban pollution, water withdrawals for irrigation and other consumptive uses, and the dams. Grand Coulee, Hells Canyon, and Pelton dams block roughly one-third of the habitat once accessible to migrating salmon, effectively restricting to U.S. waters fish that had once migrated far into Canada. More than 100 other dams in the Columbia drainage, built for a variety of purposes, have altered the flow, timing, and biological character of its rivers and streams. In 1987 the Northwest Power Planning Council attributed 80 percent of the damage to fish runs to the power system.

Artificial Production

The salmon that swim the river today differ fundamentally from those of the aboriginal Columbia (see Fig. 2–2). Roughly two-thirds of the fish returning to the mouth of the river begin their lives in hatcheries. The principal remaining species are chinook or king salmon (*Oncorhynchus tsawytscha*), steelhead (*O. mykiss*), and coho or silver salmon (*O. kisutch*). The coho, which once ranged several hundred miles up beyond the Yakima basin, are now limited to the habitat below Bonneville Dam, 60 miles from the ocean. Coho are predominantly of hatchery origin, and Columbia River silvers constitute an important component of the ocean fishery. Chinook and steelhead are upriver fish still, spawned in hatcheries but also in tributary streams and the main stem of the river. Both chinook and steelhead are prized as game fish, earning far more per fish for human economies than the commercial harvest, since sport fishers pay to fish, not to catch. The high dams—Chief Joseph on the mainstem Columbia and Hells Canyon on the Snake—deny all salmon their former range into the uppermost reaches of the river basin. Because a salmon does not eat once it leaves the ocean, the fish that traveled the greatest distance were also the largest and richest; thus it is the largest fish that were extinguished. Although technology could provide means for salmon to pass over the high dams, the insurmountable barrier lies on the other side: reservoirs so deep, with water so slow-moving, that juveniles do not have enough current to find their way out to the sea. The reservoirs and dams have also virtually

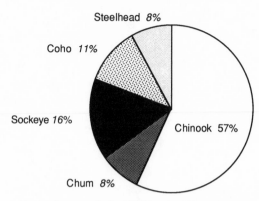

a. Predevelopment (before 1850): 11 million per year

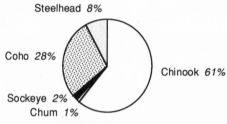

b. 1977–1981 average: 2.9 million per year

FIGURE 2−2. Species composition of Columbia River salmon. Area of circles is proportional to estimated population sizes. (Source: Northwest Power Planning Council.)

eliminated sockeye (*O. nerka*), whose juveniles require lakes; the habitable lakes in the Columbia basin are far upstream, past too many dams.

As the fish runs declined, so did the tribal civilization built around the salmon. Few cared. Indians were wards of the federal government. The fish were the province of white-owned canneries, fishing fleets, and state fish and game wardens. Their focus was economic exploitation, not ecological stability. Although these two objectives are not necessarily incompatible, the dominant economic view over the past century has drastically shifted the ground on which compatibility can be discussed.

Hatchery production has shifted both species composition and geographic origin (Fig. 2−3). Wild salmon return to their natal

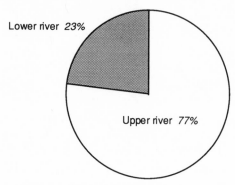

(a) Predevelopment: 11 million per year

(b) 1977–1981 average: 2.9 million per year

FIGURE 2–3. Distribution of Columbia River salmon, showing abundance above and below the site of Bonneville Dam. Area of circles is proportional to estimated population sizes. (Source: Northwest Power Planning Council.)

streams to reproduce and die. As a result of the shift of production to the lower river basin, far fewer fish now run to the upper river, beyond the Cascade Mountains. That is the region where Indian tribes' traditional fishing grounds are located; here too, there are four populations of salmon that have declined so drastically that they have been proposed for protection under the Endangered Species Act.

A driving force behind these changes has been the non-Indian harvest. In order to sustain commercial fisheries, governments and other entities have built and operated salmon hatcheries along the west coast of North America, including more than 80 in the Columbia basin. Although hatchery construction was originally authorized to mitigate damage resulting from the construction of dams, an agreement (which did not include Indian tribes) put most of the salmon production west of the Cascades, downstream from all or most of the dams. This choice avoided losses from the dams and reservoirs and allocated the bulk of the fish to non-Indian harvesters.

The decision to produce fish primarily by artificial means has also had major consequences for the wild stocks that remain. By providing a protected environment, hatcheries can enable a much larger proportion of young fish to survive until they are ready to migrate than would be the case in the wild. If this advantage persists until maturity, hatchery stocks can be harvested at higher rates than wild fish, because hatcheries require fewer eggs to produce a given number of reproducing adults in the following generation. "Surplus" adults can then be caught in ocean and river fisheries.

Hatcheries can also do considerable harm. Their close quarters permit the spread of diseases. Fish are spawned from adults selected by hatchery workers, not by natural competition, and the young are genetically uniform; thus they are more likely to succumb to disease, competition, or environmental stress after they are released. Mating of hatchery-bred stocks, which in many cases have been taken from distant wild populations, with wild fish mixes gene pools adapted to different environments, and this too can reduce the ability of hatchery fish to survive. Artificially cultured fish can spread their diseases and defects to wild populations. Moreover, capturing wild fish to stock a failing hatchery further depletes the population that would spawn naturally. Even when hatchery fish are healthy, their abundance can increase competition with wild fish for food, space, or mates, if the hatchery fish mature or migrate at times when wild stocks are also occupying a limited habitat.

Finally, the high rates at which hatchery stocks can be harvested add to the pressure to increase the permissible catch. Such increases deplete populations that cannot withstand heavy exploitation—including most wild fish. The Columbia River Inter-Tribal Fish Commission found, for example, that "the management of ocean and lower river fisheries for abundant hatchery coho runs has sharply diminished naturally spawning lower river coho." Wild coho in the lower Columbia are now extinct.

But since two-thirds of the fish are hatchery-bred, eliminating artificial production is not a practicable option. More generally, fishery managers are faced with difficult choices about which stocks to favor when salmon populations compete for habitat or wild and hatchery populations are caught together. State fisheries managers poisoned lakes in the upper Snake River drainage to eliminate sockeye salmon in favor of trout. Now, ironically, the remaining sockeye are protected by the Endangered Species Act.

Trapped but Not Domesticated

As the resources of the Columbia basin were harnessed for farming, fishing, electric power, flood control, waste disposal, municipal and industrial water supply, navigation, recreation, and other purposes, the once-inexhaustible fish runs were neglected and decimated at every stage of their wide-ranging life cycle. Spawning grounds were blocked or destroyed, migration routes made hazardous, and ocean and rivers filled with human predators. None of these dangers lay within the spectrum of natural conditions to which the fish were adapted by evolution.

The Columbia is no longer a natural river. The well-being of the ecosystem and its component species depends upon human understanding and action. Yet human management is hampered by the multiplicity of the Columbia's riches. Each of the major uses of the basin's resources is managed by a different constellation of human institutions, each set of managers guards its rights and prerogatives, and none is sufficiently powerful to bring the others to heel. Multiple management of multiple uses produces a tragedy of the commons. The salmon dwindle or perish.

The Columbia basin has been trapped rather than domesticated; it responds to human dictate, but it does not flourish. Its salmon are bred, transported, and caught under the supervision of human managers. The control exerted by those managers is limited: they cannot determine weather or ocean conditions, nor can they extirpate the diseases and animal predators that compete for salmon. But we no longer have a choice whether to manage the salmon or not; we have only the choice whether to manage well—and, if we choose, to learn how to do better over time.

Energy Economics and the Northwest Power Act

The prosperity of the Columbia basin is also tied to its dams. Like the salmon that were the mainstay of its Native American culture, the annual harvest of water from snow and rain, collected and harnessed by the dams, has become the lifeblood of the basin's industrial economy. Until 1980 it seemed that the fate of the river would remain tied to that of the power system. Then came the energy crisis in the Pacific Northwest. What happened then illustrates how ideas about natural resources can emerge in crisis. Both the new ideas and the turbulent, political way in which they emerged suggest how sustainability is likely to come about.

"Cheaper than Rainwater"

Like much of the American West, the economies of Idaho, Montana, Oregon, and Washington are greatly influenced by the federal government, whose policies have been a shaping force in regional affairs since the Great Depression. With Grand Coulee and Bonneville dams, two major projects of the 1930s, the federal government transformed the regional economy. When the development of the Columbia basin ended in the 1970s, three-quarters of the power in the Northwest came from falling water, and electricity rates were the lowest in the nation.

The New Deal in the Northwest was anchored in the Bonneville Power Administration (BPA), a federal agency created to market the enormous output of the national government's dams on the Columbia. But Bonneville soon became a regional economic force whose influence extended beyond the Federal Columbia River Power System, which generated nearly half the electricity produced in the four-state region. Although the dams themselves are operated by the U.S. Bureau of Reclamation and the Army Corps of Engineers, BPA power sales have usually governed river operations. Bonneville's influence came to rival that of state governments, and an appointment as agency head was seen by many as the region's most important political plum.

Bonneville is largely invisible to the retail consumer, marketing more than half of its power to utilities, which then supply businesses and households (see Fig. 2–4).[2] Two-fifths of BPA's sales are directly to Northwest industrial customers. Most of these are large aluminum refiners, which account for 40 percent of the nation's aluminum production capacity.

When Bonneville was not invisible it was benevolent. Power supplied by the federal dams was cheap because of federal financing,

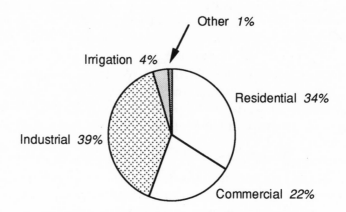

Total = 16,621 average megawatts

FIGURE 2–4. Electricity use in the Pacific Northwest in 1988, by sector. (Source: U.S. Department of Energy 1989a, p. 20.)

the low wages of the Depression, and the large inherent capabilities of the river and the sites at which dams were built. Woody Guthrie sang of " 'lectricity runnin' all around. Cheaper than rainwater." BPA's wholesale rates did not change from 1937, when power was first delivered, until 1965; and for some time afterward it remained the least expensive power anywhere. Low cost enabled the New Deal to draw upon cheap power as a tool for social change, especially by bringing electricity, and thus the conveniences of modern life, to isolated farm families.

> Lots of folks around the country,
> Politicians and such,
> Said the old Columbia wouldn't never 'mount to much.
> . . . But with all of their figures and all their books
> Them boys didn't know their Royal Chinooks.
> . . . It's a good river.
> Just needs a couple more dozen big power dams
> Scattered up and down it.
> Keepin' folks busy.
> . . .

Well the folks need houses and stuff to eat,
And the folks need metals and the folks need wheat.
Folks need water and power dams.
And folks need people and the people need the land.

This whole big Pacific Northwest up in here ought to be run,
The way I see it,
By Electri-sigh-tee.

The big power dams did come, and the royal chinooks got scarcer.
And, by the 1970s, so did the electri-sigh-tee.

The Northwest Power Act

From the turn of the century, when steam-driven turbines came into service, until 1970, electric power was a story of increasing scale and declining cost. Growth in demand brought decreasing cost per unit sold: growth benefited producer and consumer alike. In the 1970s, however, the benefits of growth vanished as costs of construction, borrowed money, and environmental damage all rose ominously. In the Pacific Northwest, industry leaders and their critics turned to the federal government for solutions. The product of their political labors was the Northwest Power Act of 1980. Like most legislation, the act embodied many compromises, compromises that had unexpected consequences when the economy took a surprising turn.

The act was designed to solve a set of social problems by technological means. As demand for power grew during the 1970s, more power plants seemed necessary to utilities. The utilities proposed federal legislation to enable them to build more plants in 1977. Yet citizen activists, whose voices were growing steadily in power and influence, argued that energy conservation could meet the demand for power at lower environmental and economic cost. The search for compromise took more than two years. Toward the end of that search, the Indian tribes and fishermen who had fought over the salmon made common cause, demanding that the damage to the Columbia's fish runs be repaired. Rather than choosing among these partially conflicting claims, Congress sought to accommodate them all. The result was an increasingly complex piece of legislation, whose implementation has taken turns unanticipated by those who fashioned the compromises.

To the utilities, the major challenge was to build new generating plants to augment the hydropower. Pressed by Bonneville, publicly owned utilities launched five nuclear power plants sponsored by the Washington Public Power Supply System (WPPSS), a public utility consortium based in Washington State but drawing upon the credit-worthiness of more than 100 utilities throughout the Northwest. Legislation intended to buttress these arrangements came too late. Plagued by cost overruns, high interest rates, and swiftly falling demand for power in the 1980s, WPPSS completed one plant, moth-balled two, and canceled the other two. That action put at risk $2.25 billion in municipal bonds, money borrowed to construct the two canceled plants. In July 1983 WPPSS and the utilities behind it defaulted on the bonds, and one of the nation's largest securities lawsuits was launched. The suit was finally settled out of court in 1989; approximately $700 million of the original $2.25 billion was repaid to bondholders and their attorneys.

Even as Congress hurried to preserve the low-cost power of the Northwest, the ground was slipping out from under the utilities. The costs of new power plants began to come due,[3] and rates increased rapidly. The rise was all the more dramatic because of the low historical base from which it started. From 1979 to 1984 the Bonneville wholesale rate increased more than 700 percent, and the retail price of power followed, more than doubling on average. At the same time high interest rates, together with a worldwide economic slowdown triggered by the oil crisis of 1979, depressed the Northwest economy, hitting its energy-intensive industries with gale force. In the rural hinterlands, layoffs and skyrocketing utility bills stirred rebellion.

By 1982, as the Northwest Power Act was in the early stages of implementation, the expected power shortage that had motivated its enactment had evaporated. Demand was far below expectations because the economy was in recession. With rates rising rapidly, conservation gained plausibility. Instead of a deficit, there was a surplus of power through the 1980s.[4] Instead of a financing mecha-nism to build new power plants, the Northwest Power Act became the blueprint for a laboratory of energy and environmental conserva-tion.

The claims of Indian tribes posed another threat to the region's power supply and economy. After their initial victories over fish harvest, the tribes filed more cases. Rulings in the lower courts suggested that the tribes might be awarded a right to enjoy a productive natural system, in which there would be enough fish to assure a reasonable standard of living. No one knew what such a decision would mean in practice, but its implications were clearly large: salmon runs had suffered major declines. Reversing that history would be costly even if it were feasible.

The Planning Council

Congress passed the Northwest Power Act at the end of 1980, as Jimmy Carter, the environmentalist president who tried in vain to create a durable national energy policy, prepared to leave office. The government could legislate and tax, but it could not make kilowatts or fish. A law could not solve the problems of salmon or power, but it could arrange for their solution over time. The power act used a familiar strategy of governance, defining a new process so that an array of choices could be made without further appeals to Congress or the courts. So far the strategy has been successful: after an initial flurry of litigation over the meaning of the act for power sales contracts and other legal matters, judicial activity ceased for nearly a decade; and although petitions were filed in 1990 under the Endangered Species Act, starting a chain of events that could affect the Northwest Power Act's fish and wildlife provisions, there is little prospect of new legislation.

The centerpiece of the new process is the Northwest Power Planning Council and the two plans it has promulgated—each several times—with wide public involvement. Chartered by the four Pacific Northwest states, the council is composed of two members from each state, appointed by the governor under procedures established by state law. Under some circumstances the council has the unusual authority to restrict or redirect the actions of federal agencies. The council is in effect an interstate compact, a form of government organization that shares both state and federal authority.

In the complex world of energy policy, influence can be more important than authority. The council's influence has three bases: the credibility of its members, the analyses of its staff, and its standing among the public. Organized during the crises of the early 1980s, the council began with impressive strength in all three areas.

The first council was led by Dan Evans, a three-term governor of Washington regarded as one of the most effective state chief executives of the twentieth century. When Evans was appointed to a vacancy in the U.S. Senate in 1983, I was named to fill his seat on the council. Though less well-known than Evans, the rest of the group was formidable as well. My new colleague Larry Mills, for example, was a former state legislator who had chaired the 1980 Idaho for Reagan organization. Although this group of Westerners seemed to have little in common with a native New Yorker living in Seattle who was a college professor, Larry and the other council members welcomed me with grace. The large tasks we faced afforded common ground enough.

We shared our plight with the council staff, an energetic and bright group of professionals, few of whom had previous experience in utilities or fisheries agencies. Just as trees define the ecology of a forest, the staff has defined the organizational ecology in which the council operates. Within the utility and fisheries communities—which are similar mainly in their clannishness—resentment of the council staff's disregard of tradition has been mixed with admiration for their inventive competence. These reactions, together with the degree to which other organizations hire professionals of similar style, have changed the terms in which issues are debated, affecting both the content and durability of the council's policies.

Those policies are embodied in two regional plans, described below. Both are products of extensive public discussion. Although public participation declined substantially as electric power rates stabilized in the mid-1980s, many divergent interests are at stake in the energy-dependent sectors of the economy and in the environmental management of the Columbia basin, so a broad spectrum of observers continues to pay attention. Thus the pluralist approximation to democracy—characterized by one political scientist as majority rule by the minority that cares—arguably remains valid. Nearly all observers agree that the council's plans articulate a public interest that goes beyond the claims of utilities or other interest groups.

The council's primary task is to formulate a plan to guide electric power development, including energy conservation. Three versions of the plan have been issued, the most recent one in 1991. Their central premise is regional cost-effectiveness, planning that minimizes costs across the Pacific Northwest's many utilities, a rule that the fragmented industry would not naturally follow.

The capital intensity of electricity production makes reliable predictions of demand essential. When capital is costly, building excess supply can be financially disastrous. Conversely, insufficient power can inflict unacceptable economic hardship on customers.

The rapidly changing energy economy of the years 1973–1986 made forecasting difficult, however. The first plan, issued by the council in 1983, included four forecasts, each representing a different scenario of regional economic development over the next 20 years. The use of multiple forecasts framed a planning approach that recognized the risks of electric power investments. Nearly half the time needed to build a new generating station is taken up with planning, design, and regulatory compliance, but as little as 5 percent of the cost is incurred at this stage. Once construction starts, costs mount swiftly. The council proposed that BPA's financing authority be directed to start more projects than were likely to be needed, delaying a final decision about which options to exercise until the need for them was closer and clearer. That way, if demand was high it could be met; if it was low, relatively little money would be lost.

The regional approach emphasized a portfolio perspective: the objective was not any single power plant or conservation program, but the whole set of available alternatives. Cost-effectiveness was to be the governing principle of management: the least costly options should be exercised, and costs should be reckoned on a regional, life-cycle basis. It was a businesslike and slightly idealistic gospel to preach to a utility industry riven by factional rivalries and facing political confrontation from once-docile ratepayers.

At the time, the recession-hobbled economy was not demanding much energy. Wary of issuing a plan that would languish unused, the council identified steps that needed to be taken: creating conservation programs, developing options for the regional portfolio. The lowest-cost items in the portfolio turned out to be conservation.

Energy Efficiency

Although the regional-scale plan takes a centralized perspective, its key innovation—the development of energy efficiency as a source of power—is radically decentralized. Conservation entails changes in structures, appliances, motors, and lighting equipment, and a host of other shifts in technology and procedure among a good share of the 9 million ratepayers of the Pacific Northwest. Bringing about these millions of changes was crucial to the development of conservation as a resource. Owners of buildings, for example, had to be

motivated to participate, structures audited for conservation opportunities, contractors and financing assembled to install conservation
measures, and monitoring of savings and collecting of costs conducted for some time after installation. What was done for buildings
had to be extended to commercial equipment such as lighting and
refrigeration, industrial processes that use electricity, and other uses
such as street lighting by city governments. None of this fell within
the traditional scope of electric utility functions; all of it needed to be
organized.

Conservation made economic sense: because electric energy had
been cheap for nearly half a century in the Pacific Northwest, power
was being used more freely than was warranted now that rates were
higher. Under these conditions, existing buildings and technologies
could be retrofitted relatively economically and new ways of doing
things instituted to save money both for customers, who would see
their bills decline, and for utilities, who would be able to defer
building costly generating facilities.[5]

Would utilities and ratepayers do what was economically rational
even though there was no power shortage? The council called for
rapid action in 1983, arguing that since conservation was a largely
untried method of obtaining power resources, it was imperative to
build the capability to develop it. Building that capability was assigned to BPA and the utilities as their top priority. Both results and
expenditures were impressive for a period of surplus: long-term
demand reductions of nearly 300 megawatts—enough conservation
to eliminate the need for a medium-sized coal plant—at a cost of
$600 million—about half as much as a coal plant of 300 megawatts
would have cost. Conservation could be obtained, and at a cost
much lower than the generating plant it displaced.

Because of its dispersed nature, conservation poses the problem of
lost opportunity: the need to affect energy users' behavior at a time
when they may not be paying attention to operating costs. Conservation is mostly investment—in better-insulated windows, more
efficient motors, and the like. The investments are made, however,
by firms and households rather than by utilities. Conservation cannot be sized or scheduled readily to meet utility needs. Worse, the
power system can easily misjudge the targeting, intensity, or kind of
incentives needed to recruit investors. For example, fewer than 10
percent of the electrically heated houses built in the Pacific North-

west in the past decade include all the cost-effective efficiency measures. The other 90-odd percent are a lost opportunity: their remaining potential for efficiency can be only partly recaptured by later
retrofitting; even in houses in which subsequent action is technolog-
ically feasible, it may no longer be cost-effective.

Hood River. Although the problem of lost opportunities was recognized in the original power plan, it remains difficult to affect the
behavior of large numbers of people who are ignorant of and indif-
ferent to electric utilities. Responding to a suggestion from the
environmentalist Natural Resources Defense Council, BPA and two
utilities attempted to market energy efficiency under laboratory conditions. The locale chosen was Hood River, Oregon, a small city
located on the banks of the Columbia where the river cuts a scenic
gorge through the Cascade Mountains east of Portland. Hood River
County is served in part by Pacific Power, the region's largest
investor-owned firm, and in part by an electric cooperative with
access to Bonneville's low-cost federal hydropower. The town has
both urban and rural consumers, and its weather is a mix of the mild
marine climate west of the Cascades and the colder winters of the
interior.

The three-year, $20-million project was highly successful in reach-
ing energy consumers but less so in saving energy.[6] With advance
research followed by a full-scale marketing effort, the project per-
suaded the owners of 85 percent of the residential structures to
participate, retrofitting just under 3,000 buildings. In participating
houses, 83 percent of the efficiency measures recommended were
adopted. These high figures reflect both the fact that the program
required no payments from homeowners, and the intensity of the
marketing effort.

An analysis of energy use found that savings were lower than
anticipated. Net savings per house averaged 2,300 kilowatt-hours in
1985–86, after conservation measures were taken. The project cost-
effectiveness limit was set at $2,645 per house, but actual expendi-
tures averaged $4,820 per house—an 82 percent cost overrun.[7]

The difference between expected and actual demand reductions
seems worrying. Yet the Hood River project had the characteristics
of both a research effort and a social program. Conservation was a
new and unfamiliar activity, not a well-tried engineering option.

Monitoring of energy uses in about 10 percent of the houses weatherized in Hood River, together with surveys, statistical analyses, and field observation of the social dynamics, cost $5.6 million, about half the cost of implementing the energy-efficiency measures. The evaluation lasted three times as long as the project itself. These ratios are typical of social experimentation, and the expenditures supported well-designed research that produced important insights into and documentation of the social process that lies behind using energy more wisely.

Perhaps the chief lesson learned was the critical importance of people and organization. Managerial flexibility was the cardinal virtue: the project needed to be revised repeatedly as the complications of acting in a real community with real contractors became apparent. Project staff discovered that they were reinventing some wheels. Retrofitting windows and installing insulation are essentially remodeling activities, and hiring a general contractor would have cut the administrative workload substantially. It was also important to enlist cooperation by people of divergent interests. An oversight committee, with members from environmental groups and utilities, wrestled with problems of implementation, hammered out compromises—and put aside their differences to act as advocates for the project when budgets or external difficulties needed attention.

Conservation challenge. Hood River established a crucial principle: Energy efficiency can be bought and paid for as a *supply* of energy— that is, on a basis similar to that of generation. Not using power is an economically sensible option to exercise. Sustainability can be part of normal business practice. But complications remain.

Bonneville's conservation staff tried repeatedly through the 1980s to design and carry out an effective conservation effort involving commercial structures. By its own admission, "None of these efforts proved able to capture significant amounts of savings." In Seattle, which counted on commercial-sector conservation to achieve overall levels of savings mandated by the council's plan, performance levels were only half what had been expected.

Conservation expenditures must generally be made during construction or renovation, a time when keeping future operating costs low may not seem important. Energy-efficient technology generally costs more to produce and install, so its cost-effectiveness depends on how long it is used. But commercial spaces are renovated in response to short-term market forces. Moreover, commercial firms operate under leases, franchises, and administrative arrangements that frequently separate financial responsibility from operational control of energy use. These difficulties of capturing lost opportunities all highlight the advantages of a direct but politically infeasible method, raising rates, which presumably would induce compliance through consumers' responding to rising bills. Using utility and government programs to anticipate and forestall higher rates can be a lot less effective technically, though much more acceptable politically.

A different problem is presented by lost opportunities in new buildings. Efficient building technology faces two hurdles. First, if energy prices are increasing, the structure should be built to be cost-effective at the expected cost of energy, not at its current cost. Second, builders are often different from owners, and owners from occupants—and only the last benefit directly. These facts indicate that building codes should exist and be enforced.

Building codes, however, are enacted by governments sensitive to political pressures, especially those exerted by well-organized constituencies. Builders of residential housing are formidably well-organized. They are also responsive to the leadership of mass-market builders, who aim at first-time home buyers. These are families just barely qualifying for a mortgage, for whom any increase in the price of the home may be unaffordable. Builders claiming to represent these aspirants to the American dream have thwarted adoption of statewide codes in Idaho and Montana, and they delayed upgrading of codes in Washington and Oregon for eight years.

During a time of power surplus, utilities also remain ambivalent about conservation, because it has the immediate effect of lowering revenues when demand is already slack.[8] In a comprehensive evaluation, the council's staff found that except in the Bonneville and Puget Power systems, few conservation programs were in place. Puget Power, whose service area includes the rapidly growing Seattle suburbs, was the only Northwest utility needing new, costly power; thus it had an incentive to develop conservation as a way to slow politically unpopular rate increases. With its surplus, BPA's commitment to conservation is also politically rather than economically motivated—a noteworthy instance of government efficiency.

The conservation efforts of Pacific Northwest utilities and governments amounts to a proof of principle. It is possible to obtain conservation savings at costs that compete favorably with new generation, and to do so within a highly fragmented utility industry and governmental structure. Achieving all cost-effective savings remains impossible for now, however, because of the difficulties of matching program action to the diverse incentives facing energy users.

Bringing Back the Salmon

The Northwest Power Act had two principles to prove: that energy conservation made good business sense, and that the Columbia's salmon runs could be salvaged while preserving the dams and their economic benefits. The act articulated a policy that is easily stated but difficult to define and pursue: electric power consumers are obliged to fund, through the Bonneville Power Administration, a program "to protect, mitigate, and enhance fish and wildlife to the extent affected by the development and operation of any hydroelectric project of the Columbia River and its tributaries." The word *any* reinforces the injunction to treat the Columbia River and its tributaries as a system. The Northwest Power Act is about an ecosystem, not only the species it harbors.

The Columbia Basin Program

That injunction has produced a wide-ranging effort, specified in the program created by the Power Planning Council. It has been a systematic but not a coordinated attack. The design calls for action at all the principal points at which human actions affect the fate of salmon: during harvest, at hatcheries and spawning grounds, and along migration routes in the mainstem river. But the limited authority conferred by Congress on the council has made coordination of the rehabilitation effort indirect at best.

The Northwest Power Act limits the contribution of electric ratepayers to damages attributable to hydroelectric power generation. In 1986 the council drew upon anthropological and historical studies of the wilderness Columbia, together with legal analyses of the industrial river, and set the responsibility of present-day ratepayers at between 8 and 11 million adult fish per year. The loss of this many fish, above and beyond the remaining 2.5 million returning to the river, could be ascribed to hydroelectric power generation. The practical meaning of this responsibility is unclear, however, since the biological capability of the remaining habitat and technically feasible hatchery sites may fall well below 8 million fish.

More specific guidance is formulated in the council's Columbia River Basin Fish and Wildlife Program, including an "interim" goal of doubling salmon populations over an unspecified time. This goal is "a signal that the program is a long-term, serious effort to solve complex problems not amenable to quick-fix remedies." Implicit in the word *longterm* is times of *biological* significance—several generations of salmon, which live from four to six years. These times are long by comparison with the terms of office of public officials.

Doubling populations while continuing to harvest at levels similar to those of recent years increases cost and biological risks: large-scale reliance on hatcheries is unavoidable. The program emphasizes learning and sustainability so as to limit cost and risk. Rapid learning—including effective evaluation—lowers costs. Aiming at sustainable increases in fish populations implies practices that lower risks to salmon gene pools.[9] And avoiding an explicit deadline to achieve doubling permits time for learning how to rebuild fish populations in a biologically sound and economically sensible way.

The council program identifies measures to be taken by the Indian tribes and government agencies that exercise management responsibility for fish and wildlife, by hydro project operators and regulators, and by those charged with land and water management. Most of the program measures are funded by BPA, at a total annual cost of about $130 million. This is about 7 percent of the agency's $2 billion annual revenues, comparable to Bonneville's most recent rate increase. Though modest from that perspective, the program's funding is unprecedented for fish and wildlife, as is its spatial scope, overlapping four states and covering a land area the size of France. About $50 million is spent directly by BPA through contracts with the Indian tribes and fish and wildlife agencies. The remainder of the cost is incurred as lost revenues—earnings forgone because water is released to benefit fish rather than power users.

The Columbia basin program is also ambitious in its foray into land and water management. Because of the planning council's energy responsibilities, the question of hydro development on small streams fell within its jurisdiction. In 1988 the council adopted a policy on "protected areas"—stream drainages where hydropower was restricted in order to protect salmon. Forty thousand stream miles were included in these areas. In this way, a major use of land and water in the still-wild Northwest was limited—although the council did not have explicit authority to control either land use or water rights.

Harvest controls. The Northwest Power Act's fish and wildlife language responded to the crisis in fisheries management brought about by litigation over Indian treaty rights. The sudden imposition of a requirement to share the catch equally between Indian and non-Indian harvesters forced the creation of a new set of institutional mechanisms to regulate fisheries. As the new rules have taken hold, there have been major reductions in catch. For example, ocean harvest levels of coho salmon have fallen by more than two-thirds from the peak year of 1976. Although landings in Indian fisheries have increased as the tribes have moved toward their legal entitlement of half the harvest, total catch of Columbia River stocks is substantially below the levels recorded in the years preceding adoption of the Northwest Power Act.

Because harvest was the central concern, and because the Indian tribes and fisheries management agencies were well-organized and vocal, the policies in the act and the resulting Columbia basin program were aimed at rebuilding a *harvestable* population of salmon—a goal that requires hatcheries. As doubts concerning the biological soundness of hatcheries emerged in the council planning process, increasing emphasis was placed on assessing conflicts between wild stocks and artificially produced fish, and on a systemwide approach that included harvest as a factor to be weighed explicitly in managing fish production. The council continued, however, to defer to the authority of the fisheries management agencies and Indian tribes on matters concerning harvest.

System planning. Although the Northwest Power Act gave a leading role to the Indian tribes and fish and wildlife agencies, they had been the principal combatants in the bitter fight over treaty fishing rights. By 1982, when the first Columbia basin program was adopted, these two groups were still living under a fragile truce. As Indian and non-Indian resource managers cautiously felt their way toward shared jurisdiction in the early 1980s, their decisions were made by consensus. Accordingly, their proposals to the council simply stapled together earlier requests and set no priorities. The result was undigestible, both to utilities wary that the program might rapidly turn into a multimillion-dollar giveaway, and to a council seeking to fashion rational responses to contentious issues.

The council asserted a shaping role, initiating in 1987 a continuing effort called system planning. System planning involved trying to think about the interactions among the hundreds of activities affecting the abundance and health of the basin's fish and wildlife, including changes outside the scope of the program. The policies established to guide system planning elevated the importance of conserving the genetic heritage of the basin's salmon, reflecting a commitment to long-run sustainability of the rejuvenated populations. System planning sought to govern the Columbia drainage as an ecosystem, with neither the authority nor resources to enforce its writ. To a surprising degree, it has worked.

Strategies for rebuilding salmon runs usually redistribute them. Before the Indian treaty litigation, the drift had been downstream, but the reassertion of treaty fishing rights focused attention on rehabilitating fish runs by bringing them back to the upper river. The principal current strategy is called *supplementation*, a technique of releasing hatchery-bred juveniles into underpopulated streams before the fish migrate to sea. Supplementation is thought to combine some of the advantages of hatchery culture, especially the high survival rate of young fish before they are released, with some of the strengths of natural production, because the selective forces that determine which fish survive after release are largely natural. Supplementation thus promises effective use of existing and new hatchery capacity, together with the hope of rebuilding wild stocks in their native streams and at population levels that will permit harvest.

Supplementation may not work. It has never been tried on as large a scale as is being proposed in the Columbia basin. Even if it does work, there may be adverse effects on wild stocks. Because it is intended to create naturally reproducing stocks that mimic those in the wild, supplementation deliberately increases competition by

introducing artificially reared fish, and contemplates interbreeding
between wild and introduced fish.

River operations. The most vexatious and economically important elements of the Columbia basin program are those relating to changes
in mainstem river operations. The river habitat is crucial for all
salmon because they must traverse it twice: once as juveniles migrating to the ocean, and later as adults returning to reproduce. The
dams pose two major hazards. First, the physical barriers themselves
cause high mortality among juvenile fish passing through generating
turbines. Second, impoundment has changed the river into a string
of lakes. Migrating fish have lost a strong current to carry and guide
them, and the reservoirs harbor predators different from the ones
the salmon are adapted to resist.

In 1990 petitions were filed on five stocks of Columbia River
salmon, under the terms of the Endangered Species Act. These
petitions followed several years of intense controversy in the Northwest over the Northern spotted owl (*strix occidentalis*), a bird
thought to be endangered by logging operations in the old-growth
forests of Washington, Oregon, and northern California. As in the
case of the owl, the salmon petitions raised the specter of severe
economic dislocations, since the federal government's duty to protect an endangered species supersedes its role as operator of economically important facilities such as irrigation systems and hydroelectric
power dams. It seemed possible that the battle over the salmon of
the Columbia would lead to drastic modifications of the way the
river was operated.

The Northwest Power Act specifies that the Columbia basin program will "provide flows ... between [hydroelectric] facilities to
improve production, migration, and survival" of anadromous fish.
From the outset it was unclear how to ground such a policy objec-
tively, because the biological benefits were difficult to assess precisely
in the massively altered ecosystem. As legal scholar Michael Blumm
commented in 1982, "the fish and wildlife agencies and Indian tribes
are unlikely to be able to prove on the basis of demonstrated factual
evidence that the mainstem flows they seek will result in specific run
size increases. . . . the standard necessarily must be something less
than scientific certainty." The indisputable decline of upriver fish
runs made it clear, nonetheless, that flows had to be improved.

Hazards at the dams can be minimized through the use of mechanical devices that collect and redirect migrating juveniles around the dam, avoiding the turbines. Bypass is costly, however, and appropriations must compete with other water-project funding in a deficit-plagued Congress. In the meantime, some protection can be provided by releasing some water at dams without putting it through power turbines, so that fish can pass the dam without a life-endangering toboggan ride through the powerhouse. This operation, known as spilling water, directly deprives the hydroelectric system of economic return, but spill requires no new construction and can be used where bypass protection is absent or ineffective.

Water budget. The keystone of the Columbia basin program is an augmentation of the flow of the river, called the water budget. Before the dams were built, flow was heavily concentrated in the spring, when mountain snow rapidly melts. Spring floods carried juvenile salmon to the ocean. The trip is made much longer now by the fact that water flows more slowly in reservoirs than in the unimpounded river. The slowdown exposes juvenile fish to predators for a longer time. In addition, salmon must make a physiological changeover in going from fresh to salt water; if arrival in salt water is delayed, the fish may revert to a freshwater constitution, stop growing, and never reach harvestable size. For years the fish and wildlife managers and Indian tribes had requested higher flows in the springtime migration season. But the requests carried no authority, and the dams, controlled by utilities and the Army Corps of Engineers, were run to optimize power revenues most of the time.

Led by chairman Dan Evans, the planning council conducted extensive negotiations with utility experts and tribal and state agency leaders in 1981 and created a water budget that restored part of the spring flood. The water budget is a quantity of water, approximately equal in energy capacity to a medium-sized coal plant, that can be controlled by fisheries managers. During the spring, when fish are observed to be migrating in the river, the fisheries managers call for water-budget flows to commence. Most of the time, once the augmented flows have started, they continue until the budgeted water is used up. Power managers can use water flowing in the river to generate power, but they do not control its timing. For this reason, the water budget loses money for the power system, because water is released in the spring rather than in the autumn or winter, when higher demand for power would permit higher prices. The water budget, reckoned by the revenues forgone, costs about $40 million per year on average.

The water budget is a more generous compromise for the Columbia than for the Snake, because the upper Columbia discharges more water and has substantially more storage in its upstream dams. In practice, even the water dedicated to the water budget in the Snake River drainage has often been unavailable.

The biological benefits from the water budget are hard to see, in part because of its small size in comparison to natural fluctuations. Figure 2–5 shows a compilation of data from the Snake River on the relation between the volume of river flow and average travel time for migrating juvenile fish. This handful of measurements—the black dots in the figure—constitutes the principal justification for losing $40 million per year in power revenues. Understanding what these measurements suggest and do not suggest illuminates the practical challenge of sustainable development.

Each data point represents a measurement of how fast a typical juvenile salmon moves down the river, depending on how much water the river is carrying. Like riders on an escalator, the fish should go faster when the river flows faster—that is, when the flow level is higher. Therefore, the data points should trend down to the right; the higher the flow, the shorter the travel time. The downward-pointing straight solid line represents the spring runoff in quantitative terms—the biological benefit, measured in reduced travel time, plotted against economic cost, measured in river flow. The water budget, shown as an increase in flow in Figure 2–5, should therefore produce a biological benefit.

The concept is straightforward, but there are complications. First, there are few measurements. Each point represents a sizable investment of research time and technical effort—marking, releasing, and recapturing thousands of migrating juvenile fish at dams hundreds of miles apart, to see how long it takes them to migrate downstream. At most a single data point is measured in a year, not only because of cost, but because different levels of river flow are needed to see how fish and flow relate to each other; each year, one sees no more than one Columbia—a wet one this year, perhaps, a drier one next, and so on.

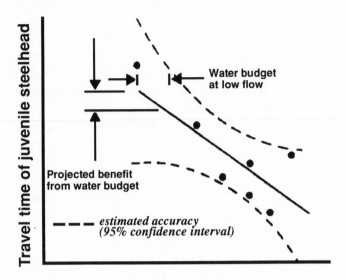

River flow

FIGURE 2—5. Effect of the Snake River water budget on juvenile salmon migration. Both vertical and horizontal scales are logarithmic. For consistency, the flow at Ice Harbor Dam is used as the standard measure of flow. The water budget in the Snake River (horizontal arrows) is superimposed on the measured relationship between river flow and migratory fish travel time. The projected reduction in travel time (vertical arrows) as a result of the water budget lies well within the measurement error (curved lines). Therefore, the effectiveness of the water budget in speeding juvenile fish downstream is difficult to prove. (Source: Northwest Power Planning Council, after Sims and Ossiander 1981.)

Second, although the relationship shows the expected trend—the higher the flow, the lower the travel time—the observations do not lie along a single line; there are fluctuations and "noise" in the data. The relationship between travel time and river flow is affected by other factors, many of them unmeasured. For example, the condition of the fish when they start migrating can make a large difference, but trying to pinpoint the health and readiness of thousands of fish the size of a human finger is too expensive and time-consuming to be practical.

Third, the uncertainty is large relative to the size of the water budget. The effect of the water budget is largest when the flow is lowest, at the left side of the graph, because the effect of a fixed volume of water is largest when the underlying flow is lowest. Yet even under these conditions, the *change* brought about by the water budget lies well within the dotted lines indicating the range of uncertainty in the available data.[10] The biological effectiveness of the water budget, even if it were fully implemented, would be difficult to observe, even in low-water years when one would expect it to be most helpful. Thus the most significant fact about the water budget may be that a substantial reallocation of resources has been made in the name of tribal tradition and environmental values.

Flaws in the System

Like most pieces of legislation, the Northwest Power Act is a product of compromise. Of the four federal agencies directly responsible for implementing the act, only the Bonneville Power Administration is assigned a comprehensive mandate to "use the Bonneville Power Administration fund and the authorities available to the Administrator . . . to protect, mitigate, and enhance fish and wildlife . . . in a manner consistent with . . . the program adopted by the Council." The Army Corps of Engineers and the Bureau of Reclamation, which are responsible for operations of federal dams, and the Federal Energy Regulatory Commission, the agency with authority to require changes in the operations of nonfederal hydroelectric dams, are directed only to exercise their responsibilities "taking into account at each relevant stage of decision-making processes, to the fullest extent practicable, the program adopted by the Council." With its clearer statutory mandate, Bonneville's actions—chiefly funding, program management, and power marketing—have been more in accord with the Columbia basin program than has been the case for the other three agencies, which control flows and spills.

Because the effort to revive declining salmon would require decades, it was appropriate to focus the effort on a systematic approach. That decision subordinated the most troubled runs, for which emergency action may be needed, to the interests of salmon throughout the basin. In 1987 a review of the fish populations suffering declines found that the most troubled stocks were ones left unprotected by fishery managers in balancing the competing needs of harvest and conservation. Yet when petitions to protect the most depressed stocks were filed under the Endangered Species Act, the fish managers joined with environmentalists in demanding that the power system provide more water to transport juvenile fish. The council was left with the task of fashioning a sensible balance.

No one acts in the natural environment without acting in the public arena. Political risks cannot be eliminated. With its limited powers to compel compliance and its significant capability to provide funding, the council accordingly developed a collaborative style. Council chairman Charles Collins observed: "A policy-making body will not flourish if it relies on written authority to make things happen. It must develop the kind of policy that is so logical, makes such good sense, that other parties participate voluntarily. It must have the kind of processes that involve others at every level so that the policy becomes, not the Council's policy, but the region's policy." Such an approach has created a systemwide program spending more than $130 million per year. No one knows whether it is enough to save the most threatened salmon runs from extinction.

A Fundamental Change

What is already clear, however, is that the Northwest Power Act has fundamentally altered the treatment of fish and wildlife in the Columbia basin. Indian treaty rights, ignored for more than a century, are widely accepted. More than 40,000 stream-miles in the basin have been put off-limits to hydropower development. Fishery resource management agencies in tribal, state, and federal governments exercise substantial influence in the policies and budgetary choices of the Bonneville Power Administration. There have been significant changes in power system operations and planning in response to the Columbia basin program. The impact of these measures can be gauged by different yardsticks: in lost power revenues of $40 to $60 million per year; or in power resources displaced or subordinated to fishery protection, the equivalent of a medium-sized coal plant.

These costs make the Columbia basin program the world's largest attempt at ecosystem rehabilitation. The investment, from one perspective, is about $50 per salmon, a remarkable figure when one realizes that none of the market value of the fish—about $30, de-pending on market conditions—goes to the ratepayers. Of course, not all of the current costs should be charged to the current popula-tion of somewhat under 3 million adult fish per year. Much of what is being done is meant to make the population grow in the future. Even though economists assessing the costs and benefits of compa-rable programs have arrived at the same approach as the one set forth in the Northwest Power Act, the sheer magnitude of the dollar figure continues to fuel debate.

As in the case of energy conservation, salmon rehabilitation has broken new ground. As also with the efforts in electric power, it is not clear that what is being done will be enough to achieve a workable, sustainable balance. The river is in an unnatural, partly managed condition. Increased flows in the spring—the remedy rec-ommended by the fisheries advocates—may or may not suffice to rescue the salmon. The mounting disillusionment with hatcheries demonstrates the fragility of our understanding and the unexpected, even perverse effects of earlier attempts to mitigate damage. Faced with the extinction of fish stocks, it is appropriate to try urgent measures. But it would be inexcusable, given the risk of failure and the cost of trying, to take these steps without explicit emphasis on learning, whatever the outcome.

The idea of rebuilding the fish and wildlife of an industrialized ecosystem is heroically optimistic—a hope that might not have occurred to anyone except those who had rehabilitated the Wil-lamette basin in Oregon or Lake Washington near Seattle. The extension of those learning experiences to the multijurisdictional, multifunctional situation of the Columbia basin—a large eco-system—requires coordinated action and learning on a new scale, explored in the next two chapters. In matters of effective gover-nance, the institutional "experiment" launched by the Northwest Power Act becomes important not only in itself but in the search for sustainable development.

Chapter 3

Compass: Adaptive Management

[The President] cannot count on turning back—yet
he cannot see his way ahead. He knows that if he is
to act, some eggs must be broken to make the
omelet, as the old saying goes. But he also knows
that an omelet cannot lay any more eggs.

—Theodore C. Sorensen, *Decision-Making in the White
House: The Olive Branch or the Arrows*

The Columbia basin experience identifies themes common to large
ecosystems: our control of nature is limited; there are opportunities
for constructive change; but to search for sustainability we need to
learn in a new way. The old way is trial and error, a slow process.
Deliberate experimentation can be much faster, though it can also be
more tangibly costly and politically risky. Experimentation using
adaptive management is one part of social learning. The other part is
politics: keeping within bounds the conflicts that inevitably arise
concerning large ecosystems far from sustainable equilibrium. In
this chapter I describe the first component, an approach to policy
that enables experimentation and learning to take place. Adaptive
management is being used in the Columbia basin, and in explaining
the concept I am also interpreting the story set out in Chapter 2.

Human Limitation

A durable lesson of the Columbia basin is the inherent inefficiency of
large systems. A hydroelectric power system takes decades to build,
thousands of people to operate, and affects enough people that its
fortunes can sometimes swing elections and sway national governments. For all these reasons, perfect efficiency remains an ideal.

Though infrequently articulated, this is conventional wisdom among people involved with large institutions.

Less often grasped is the effect these human limitations have on learning, especially social learning as it is discussed in the next two chapters. Social learning itself depends upon the concept of *bounded rationality*.

Originally developed by Herbert Simon, the idea of bounded rationality has influenced the social sciences from anthropology to artificial intelligence. Simon observed that human beings have a limited information-processing capability. Instead of considering all available alternatives when making a choice, we typically select from a restricted set. Instead of choosing the best alternative (which may not even be in the set we choose from), we typically make a *satisfactory* choice—one that is good enough, if not best. These apparently simple insights have profound implications.

One implication is that inconsistency is normal. Inconsistencies arise naturally as choices that are good enough are made from limited options, because the choices are not disciplined by either a search for all options or an identification of the best alternative. Moreover, if recognizing, resolving, and removing inconsistencies is a costly activity, inconsistencies will not be rooted out completely. Every child who has watched her parents struggle with the gap between moral principle and actual behavior knows this. Only those who believe in the rationality of human action are puzzled by it. But nearly all adults believe in rationality. We believe it pragmatically rather than dogmatically. That is, we use the assumption that others act rationally to infer their motives. And then we are surprised by their inconsistencies, usually turning our puzzlement into moralistic judgments about character.

Another implication is that large organizations can do things no single person can accomplish—not only in terms of scale, but also in terms of the *kind* of task performed. No single person could master enough skills to build a jetliner or provide the backup skills that make heart surgery, writing checks, or driving a car possible. But because many large objectives can be broken down into limited jobs, ordinary people can do the limited jobs and collectively do extraordinary things. Organizing the jobs and carrying out the coordination necessary to get them to mesh smoothly is itself a job, which we call management.

For our purposes, the most important implication is that learning in a world of bounded rationality is a costly, step-by-step search for better alternatives, in which local improvements may or may not benefit the whole. Better local management may enhance a salmon run, only to destabilize an ecosystem further. Operating within a large ecosystem, bounded rationality needs to be complemented by a system-sized perspective. The problem is that such a perspective is itself created, updated, and used by human beings—and therefore must also be bounded. A strategy for using bounded rationality to learn rapidly is deliberate experimentation, which isolates part of complex reality, makes simple changes in it, and watches for results.

These abstract ideas about mundane human realities may become clearer as they are fleshed out by the concrete examples discussed below. But the concrete instances need the abstractions to make them cohere with one another.

Experimentation as Policy

Because human understanding of nature is imperfect, human interactions with nature should be experimental. Adaptive management applies the concept of experimentation to the design and implementation of natural-resource and environmental policies. An adaptive policy is one that is designed from the outset to test clearly formulated hypotheses about the behavior of an ecosystem being changed by human use. In most cases these hypotheses are predictions about how one or more important species will respond to management actions. For example, commercial fishery regulation, monitored by a regulating authority, can readily be designed in an experimental fashion. If the policy succeeds, the hypothesis is affirmed. But if the policy fails, an adaptive design still permits learning, so that future decisions can proceed from a better base of understanding.

Adaptive management is highly advantageous when policymakers face uncertainty, as they almost always do in the environmental arena. But the adaptive approach is not free: the costs of information gathering and the political risks of having clearly identified failures are two of the barriers to its use. Moreover, because the adaptive model of learning does not take into account the limitations of learning within human organizations, specific precautions should be built into the design of policies. In simplest terms: without experimentation reliable knowledge accumulates slowly, and without reliable knowledge there can be neither social learning nor sustainable development. How much social learning can be afforded in particular times and places affects how quickly development can become sustainable.

Adaptive management was defined by an interdisciplinary team of biologists and systems analysts working in the mid-1970s at the International Institute of Applied Systems Analysis, a think tank located in an old Hapsburg castle in the town of Laxenburg, outside Vienna. Their work, published in 1978, was rooted in what their leader, Canadian ecological theorist C. S. Holling, called "a bias" that understanding how natural systems respond to human disturbance is essential to "living with the unexpected." A similar approach, but without Holling's emphasis on experimentation, was developed slightly later by Allan Savory in the American Southwest; he called it "holistic resource management." The implicit idea underlying these approaches, that humans could not and should not try to control as many natural fluctuations as industrialism seems to demand, was shared by environmentalists following Aldo Leopold in urging a change in "the role of *Homo sapiens* from conqueror of the land-community to plain member and citizen of it."

There was a parallel development among political scientists warning that humans in organizations could not be controlled as if they were machines, for "solutions to problems cannot be commanded. They must be discovered." I return to this line of thought in Chapter 6.

The adaptive approach to natural resource management has been developed primarily in western Canada and the U.S. Pacific Northwest, drawing intellectual energy from Holling's former students and colleagues at the University of British Columbia; part of that experience is discussed further in Chapter 5. The separate stream of social science has ranged more broadly, exploring redundancy as a strategy for improving the reliability and performance of organizations as well as empirical studies of high-reliability organizations. This book joins these two lines of investigation.

System Planning in the Columbia Basin

Adaptive management was first applied as an explicit policy on the ecosystem scale in the 1984 revision of the Columbia basin program, and became a guiding premise of the system planning described in Chapter 2. Experimentation as a strategy for managing a large ecosystem was a new idea in 1984—one consistent with but not explicitly called for by the Northwest Power Act. Because learning complemented the basic mission of rehabilitating the Columbia's salmon, the adaptive approach was incorporated as a component of policies aimed at rebuilding the fish runs, yielding social learning as a by-product. From the outset the policy design reflected an attempt to be idealistic about science and pragmatic about politics. Table 3–1 shows how learning was combined with policies for rebuilding salmon populations.[1]

The tension between scientific rigor and political reality can be

TABLE 3–1. Components of an adaptive approach to salmon enhancement. Individual components are discussed in Chapter 2 and in Northwest Power Planning Council (1987, p. 43).

Doubling goal. Numerical target for increased abundance, based upon study of losses due to previous exploitation.

Upper river emphasis. Targeting of actions on basis of location of historical losses and claims of those who suffered from them (Native American tribes).

Genetic risk policy. Recognition of biological properties that are fundamentally important, affected by management actions, but characterized by slow feedback.

Mainstem survival, production mix, and harvest management policies. Identification of major system components as foci for largely independent, major policy initiatives.

System integration and system planning policies. Drawing together of policy initiatives so as to identify cumulative effects that would be missed within subunits. Integration also defines resource allocation and planning as central, systemwide functions.

Adaptive management policy. Testing and evaluation of projects "wherever possible, taking into account the need for control or comparison cases for statistical validity."

Research, monitoring, and evaluation. Continuing responsibility for monitoring, linked to adaptive management.

Resident fish substitution policy. Enhancement of nonmigratory fish in areas where salmon are now excluded by dams, reflecting human choices to redesign an ecosystem rather than restoring it.

seen in the systemwide approach adopted in 1987. Rehabilitation of the salmon runs needed to take explicitly into account how production, both artificial and natural, would interact with the other major points at which human action affected the mortality of fish: passage through the dams and reservoirs, and harvest. With its attention to salmon, only one species among many in the Columbia basin and North Pacific, the system planning framework does not define an ecosystem approach at all. Moreover, the Northwest Power Act, as discussed in Chapter 2, prohibited the council from making decisions that would affect water rights. Nor did Congress authorize land management regulation or planning capability, although the protection of more than 40,000 stream-miles in the tributaries of the Columbia effectively limits a use of land and water that directly affects salmon. With this circumscribed mandate, it would have been inconsistent with law and politically foolhardy for the planning council to claim responsibility for the entire ecosystem.

Nevertheless, the Columbia basin program obviously exercises an ecosystem *influence*, owing both to the large resources at the council's disposal and to the centralized vision of the ecosystem that resulted from system integration and from the council's central role in building a data base describing the fish and wildlife of the Columbia basin. It seems unlikely that *any* government would authorize central control of a large river basin by an unelected body; nor would such authority seem wise, after the environmental experience of Eastern Europe. We are still left with the open question of how ecosystem influence might best be exercised to promote sustainable development. One element of any such strategy is the securing of reliable knowledge.

Reliable Knowledge, Uncertain Nature

Adaptive policies define experiments probing the behavior of the natural system. Experiments often bring surprises, but if resource management is recognized to be inherently uncertain, the surprises become opportunities to learn rather than failures to predict. Reliable knowledge comes from two procedures: controls and replication. A *control* matches what one is changing (the treatment) to a companion case in which that same factor is left unchanged (the control). The use of controls permits insight into whether it is the treatment that is causing the effect one sees, rather than something else such as a change in weather. *Replication* is essential because if knowledge is reliable it can be shown to work more than once; real relationships between cause and effect will show up consistently. When controls and replication are explicitly taken into account in policy design, the manager's understanding of the ecosystem can be tested against experience.

Although virtually all policy designs take into account feedback from action, the idea of using a deliberately experimental design, paying attention to the choice of controls and the statistical power needed to test hypotheses, is rarely articulated and still more rarely implemented. In 1977 a heroic effort was launched to save the endangered Kemp's Ridley sea turtle on the Gulf coast of Mexico and the United States. It had no controls. As a result, fourteen years and $4 million later, no one can tell if the program has worked. The environmental scientist who started the project has concluded that it has been a failure. But advocates of shrimp fishermen, who chafe under regulations aimed at helping the turtle, say there is no way to prove that the project has not achieved its aims. In 1992 they were seeking additional funding and expansion of the project as part of their campaign to roll back regulation. The value of an experimental approach is neither theoretical nor obscure.

Interventions into an ecosystem can provide insights into its behavior. Often these insights are useful because they lead to a clearer picture of the management objective, such as the relation between diversity of species and the stability of the ecosystem they inhabit. This kind of qualitative knowledge is especially useful in the absence of quantitative accuracy or predictive capability because it enables analysts to form an intuitive concept of the system. Seeing the ecosystem as a whole must precede efforts to manage it. The adaptive perspective begins from a scientific viewpoint, and its progress into the realm of action is informed more by the observational interest of the naturalist or astronomer than by the manipulative tendencies of the engineer or entrepreneur.

The adaptive approach is needed if scientific uncertainty is not to thwart socially timely action. Sustainability turns on the ability to manage large ecosystems, and thus requires understanding their behavior in practically useful ways—that is, while recovery from overexploitation is still within practical reach. But the analysis of large ecosystems faces imposing obstacles:

- *Data are sparse.* It is difficult to observe the state of the ecological system and the human economy interacting with it. Measurements of the natural world, such as the size of migrating populations, are inexact; and natural systems often yield only one data point per year (such as the number of fish in a run).

- *Theory is limited.* Reliable observations are few, and theories of natural environments do not permit deductive logic to extrapolate very far from experience. Also, human perturbations of the natural environment are frequently both large and unprecedented so that it is unclear what theory is applicable.

- *Surprise is unexceptional.* With limited theory comes poor knowledge of the limitations of theory. Predictions are often wrong, expectations unfulfilled, and warnings hollow.

Because the behavior of ecosystems cannot be understood, much less predicted, on the basis of studies done at the laboratory scale, it is essential to learn from large-scale interventions into populations and landscapes. Uncertainty makes errors and surprises inevitable; experimentation is an effective strategy for sensing surprise and recovering from error. Experimentation produces unambiguous results only when the interventions are large enough to bring about a measurable change. Large changes are often expensive, risky, and controversial, and therefore require negotiation and planning. These facts should not be taken to mean that adaptive interventions must all be systemwide. Indeed, an important skill in learning about large ecosystems is to find those interventions that can teach a lot quickly from small-scale experiments.

Adaptive management offers the hope that by learning from experience we can reach and maintain a managed equilibrium, with a resilience able to withstand surprises. The cost of experimentation on the ecosystem scale usually requires creative use of interventions made for other purposes; the Columbia basin program is a rare exception, since its goal is sustainable rehabilitation of fish and wildlife, an objective supported by a portion of the revenues earned from the sale of hydropower.

Seeing the System: Data Base and Model

The concept of ecosystem management, with or without experiments, has a logical requirement: that one be able to see the ecosystem as a whole in some fashion. This requires information, together with an analytic capability that can select from the information useful portraits of what is being managed. Until the Northwest

Power Planning Council began to assemble a systemwide data base to put already-collected information into a structure such as Figure 3–1, no single collection of information for the Columbia basin existed. This fact seems extraordinary until we also realize that much less knowledge is required to exploit an ecosystem than to try to manage it. The economic and organizational implications of the shift are accordingly large.

An analogue is television news, increasingly the primary way citizens of postindustrial societies learn about the external world. Each day's news programs are produced by large teams of journalists and production crews, gathering information from various places in the world, subjecting it to processing that ranges from developing film to preparing commentary, and then presenting it in a form that is understandable to young parents trying to keep food from spilling off the dinner table while they watch the latest developments in the Middle East, the National Cancer Institute, and Fenway Park.

Managing large ecosystems can be somewhat more orderly, and the depth of coverage considerably greater, but the essentials—

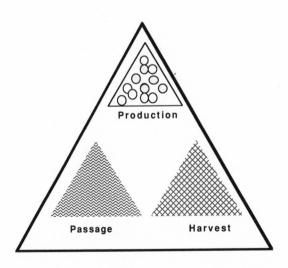

FIGURE 3–1. Columbia basin system planning framework. (Source: Northwest Power Planning Council.)

information gathering and processing capability, and an institutional and economic structure that can operate those technologies— are the same. In an experimental approach, management interacts with the ecosystem, so the links become conceptually more intricate; similarly, the purveyors of the evening news are keenly aware of the shifting tastes of their viewers, and make adjustments accordingly.

Model of the system. The processing capability requires a *model* of the ecosystem. Figure 3–1 is such a model: it describes three points at which it is convenient to sense the abundance and health of salmon, since measurements are being made routinely already; and it assumes that salmon are hatched, go to sea, grow large enough to be caught in a fishery, and return to the rivers to spawn. All along the way, the population that started as eggs gets smaller.

A model this simple seems both incontrovertible and uninformative, because our expectation is that the complexity of natural behavior emerges from complex rules. It may not. There is so much spatial and temporal variation in nature that even simple rules lead to unexpected outcomes. The last dams on the Snake River were built in the 1970s, a time when there was already appreciation of the high value of salmon, especially the prized steelhead native to Idaho's streams. Because there were fish ladders that adult salmon could climb in their homeward migration, it was believed the dams would not have a large effect. No one realized at the time that the presence of nine dams from Idaho to the Pacific would drastically reduce the juvenile salmon's chances of survival because the risk of mortality is *compounded*, like interest on a credit card loan. A fish that survived one dam still had to run the risks of all the others downstream. If each dam allowed 85 percent of the fish passing it to survive, then of 100 juvenile salmon starting in Idaho only 23 would reach Portland, Oregon. This compounding effect was "obvious" once realized. But by then the concrete was in the river, and there was no protection designed, let alone installed, for the young fish.

A model might have caught this error, simply because putting numbers together in a model requires enough careful thought that relationships that are logically linked tend to emerge. The question for the future is what errors can be caught early enough to save important biological assets or to make saving them affordable to humans.

Variability. The usual experience with ecosystem data is that there is not enough to define the biology with any confidence, but far too much for a single human mind to assimilate. Models are indispensable simply to do routine bookkeeping on large quantities of data. There are two common strategies for exploring the confusing information available. One is *comparison* across time or space or among biological parameters such as species or gender. Another approach is *simulation*, exploring the structure of the assumptions built into the model's mathematical dynamics. Simulation does not involve prediction of a determinate future, but rather use of the model to delineate "what if" cases.

Think of the data as bits of metal and stone, which are organized into a sculpture by the conceptual structure of the model. Simulations and comparisons are ways to see the sculpture from different angles, ways to walk around the data so as to appreciate the model's way of organizing parts into a whole. As data on the Columbia basin were being assembled, the Northwest Power Planning Council made a "discovery": reopening salmon habitats high up in the river would be futile because the young fish from these areas would not get to sea until dam-related mortality was reduced. Again, this was obvious in retrospect. It was not a discovery in the sense that new species are discovered—by observation—or in the sense that Darwin discovered natural selection—by theorizing. It was discovery in the sense that the assumptions contained a surprising implication.

Indispensable and always wrong. The behavior of natural systems is incompletely understood. Predictions of behavior are accordingly incomplete and often incorrect. These facts do not decrease the value of models, but they do make it clear that ecosystem models are not at all like engineering models of bridges or oil refineries. Models of those kinds of physical system often are complete enough and accurate enough to permit decisions to be made on the basis of their quantitative projections. Models of natural systems are rarely that precise or reliable. Their usefulness comes from their ability to pursue the assumptions made by humans—assumptions with qualitative implications that human perception cannot always detect.

Models and their associated data bases also provide a consistent framework for comparing alternative courses of action. "Consistent" is not the same thing as "objective." All management is biased, and the purpose here is management. What these comparisons provide is a clearer understanding of how the assumed behavior and data fit together—what I described above as "appreciation" of the "sculpture." When the analysis works, the manager develops an awareness of instabilities and vulnerabilities in the system as an interacting whole. Such insights may be called *system knowledge*, the product of probing the available data with assumptions about behavior. From this point of view, adaptive management is a means of improving system knowledge, by testing assumptions against experience.

The use of the model for real decisions has another implication. The model-using manager must provide an intellectual "paper trail," a way of understanding the chain of reasoning that leads from data base to output. The paper trail is crucial if learning is to be possible: without an understanding of how one's model of reality works, it is impossible to go back and improve that understanding when reality fails to agree with prediction.

Finally, the process of building a model is a way of working out a shared view of what is being managed and how the managing should be done. Often that process is conducted by a diverse group of people drawn from different institutions, some of them organizations with conflicting interests, such as Indian tribes and electric utilities. When this happens, model building becomes a way of negotiating.

Ecosystem models are always wrong, in the sense that reality conforms to their numerical projections only very rarely. Models are indispensable because without them human misunderstanding persists, unaware of its errors.

Managing Adaptively

Several characteristics of adaptive management must be taken into account in design and practice, if learning is to continue for the time and spatial scales needed to accumulate understanding.

- As a decision-making perspective, adaptive management is ecosystemic rather than jurisdictional. The adaptive approach crosses boundaries and links functions such as fisheries and land management.

- What is being managed is a population or ecosystem, not individual organisms or projects. Failures at the individual level need to be tolerated because risk-taking is needed if hypotheses are to be advanced and tested.

• The time scale of adaptive management is the biological generation rather than the business cycle, the electoral term of office, or the budget process.

These characteristics have suggested to some that adaptive management is ponderous and slow to act. Just the opposite should be true. The adaptive approach favors action, since experience is the key to learning. Moreover, the adaptive approach does not aim for a fixed end point: "Environmental quality is not achieved by eliminating change," Holling observed. The goal, instead, is resilience in the face of surprise. Surprise can be counted on. Resilience comes from constant testing, that is, from change and stress, from survival of the fittest in a turbulent environment.

Favorable Institutional Factors

Adaptive management makes sense, but it was adopted by the planning council more for institutional reasons than for logical ones. The institutional circumstances operating in the Columbia are listed in Table 3–2, stated in terms independent of the specific setting. The adaptive concept provided a way to grapple with the central task of the Columbia basin program: the allocation of unprecedented

TABLE 3–2. Institutional conditions favoring adaptive management.

There is a mandate to take action in the face of uncertainty.

Decision makers are aware that they are experimenting anyway.

Decision makers care about improving outcomes over biological time scales.

Preservation of pristine environments is no longer an option, and human intervention cannot produce desired outcomes predictably.

Resources are sufficient to measure ecosystem-scale behavior.

Theory, models, and field methods are available to estimate and infer ecosystem-scale behavior.

Hypotheses can be formulated.

Organizational culture encourages learning from experience.

There is sufficient stability to measure long-term outcomes; institutional patience is essential.

---

funding to the rehabilitation of a river drainage. More important, perhaps, adaptive management did not require the council to risk its future on unreliable predictions or to constrain its political discretion to conform to an academic ideal of rationality. One reason the adaptive approach is of crucial importance to sustainable development is that it leaves choices with its implementers. This discretion is the key to the political feasibility of an adaptive policy.

The Columbia River Basin Fish and Wildlife Program was adopted in haste in 1982 because of the urgent need to move on to the regional power plan and the fiscal crisis of nuclear energy. Aware of its haste, the council resolved to reexamine the fish and wildlife program by opening a formal amendment process in 1984. Fresh from its political and technical successes in the water budget and energy planning, the council wanted to put its stamp of politically astute technocracy on its efforts in behalf of salmon. Yet the vexed history of salmon management made it plain that this would be no easy task politically.

The solution was to combine the adaptive concept with a goal-setting effort that deliberately and deliberatively involved the fisheries managers, electric utilities, and hydro project operators. This process led to the 1987 version of the Columbia basin program. Setting goals for a land area the size of France is a complex affair. In this task, the idea of learning via adaptive management was both a guiding theme and a separate goal; that is, learning could be made a basic premise of implementation, while questions such as overall levels of spending and the spatial distribution of projects could be left to the political process. In these respects adaptive management is a policy with inherent institutional strengths.

Two other attributes of the Columbia basin program are less comforting: economics and transferability. The necessity of committing budgetary resources in the face of uncertainty poses an obstacle to sustainable management of most natural resources—a barrier lowered by the Northwest Power Act. By funding the Columbia basin program from the large revenues generated by the sale of hydropower, while providing only general guidance to the council on costs, the act assured plentiful financial backing. Indeed, the confusions in the original fish and wildlife program of 1982 derived in part from the perception that Congress had intended generous funding to the fisheries managers, a perception that prompted the utilities to resist, and both sides to exert political pressure. The goal-setting process and adaptive management deflected political energies away from the wasteful dynamics of the pork barrel and toward biological effectiveness. But the question of how to spend unprecedented sums arose only because of national actions that presented decision makers with a *fait accompli*. This is not a situation that is likely to arise frequently.

Overall, the conditions in Table 3–2 are ambivalent. They make clear how the Columbia basin experience could demonstrate the viability of *adopting* adaptive management as a strategy for sustainable development. But they also suggest the difficulty of using the idea in places where stable funding cannot be assured, and of perpetuating learning anywhere. To grasp the latter difficulty better, we need to understand what is at stake in taking the scientific method seriously when there is no laboratory in which to work.

Avoidable Error and Genuine Surprise

What does it mean to be prudent when there is uncertainty? First, recognize the possibility of surprise. Second, plan and act to detect and to correct avoidable error. Most discussions of rational choice begin with the second step, so it is worth spending a moment to reflect on the first.

External events and consequences of action can be grouped into three broad classes. Consider a horse race. Some events are expected: some horses will finish ahead of others. Other events are the result of random or unpredictable processes: it is often unknown *which* horse will finish first. Still other events are surprises, unpredicted and unexpected: an earthquake could bring down the grandstand during the race. Risk analysis examines, sometimes in considerable detail, the first two classes of events. But by definition, it is difficult to describe or to estimate the probability of the third class, surprises.

Adaptive management in large ecosystems is alert to surprise. Because adaptive management treats the system experimentally, the possibility of surprising outcomes is recognized from the outset. Thinking in terms of experiments has an important sociological consequence: surprising results are legitimate, rather than signs of failure, in an experimental framework. Unexpected responses look "wrong" and are always under suspicion as errors in the way the measurements were taken or as signs of mischief or incompetence by the experimenters. This suspicion is healthy and appropriate; most systems of measurement *are* vulnerable to mistakes, and most deviant results are, on closer inspection, the result of procedural errors and not real surprises in the behavior of the system. Experimentation helps to create a social system able to recognize the needle of real surprise in the haystack of mundane error.

Experimentation is also expensive, for two reasons. First, there is the likelihood of false alarms. An experimental system is not invulnerable to errors; rather, it uses a systematic method to separate errors from reliable findings. Surprising results that survive simple checks for elementary errors can then be tested—first by repeating the experiment and second by designing more elaborate experiments aimed at sorting out alternative explanations. The problem is to judge when it is worth pursuing surprises by doing additional experiments. Each experimental measurement is likely to be costly in terms of time, to require manipulation of large systems (such as a hydropower dam), and to demand scarce technical expertise. Deciding which trails to follow and which to set aside as "false" alarms is a matter of judgment. When the costs are high, these judgments are likely to be controversial.

A second element of cost is instrumentation. If the system is well understood, it is often possible to monitor its behavior cheaply. A business firm is judged by its profitability: one number per quarter can be enough for the shareholder to decide whether to hold a stock or sell it. Although firms are complex social systems, for the purposes of the investor they can treated as simple. When ecosystems are treated as simple objects, tragedies of the commons have ensued; treating common-property resources like urban air or flowing water as a simple commodity creates conditions that foster overuse of those resources. But to treat ecosystems as objects of ineffable complexity would imply giving up on management altogether. A sensible middle course is to pay the costs of sensing dimensions that are already known to be important, such as the abundance of highly valued species like salmon, while maintaining social and institutional receptivity to surprises, such as the decline of threatened fish runs, so that the investment in information gathering and analysis can be expanded in response to the unanticipated. This sounds like a platitude—until we confront the difficulty of obtaining estimates for apparently simple ecosystem variables such as the area of tropical forest cleared in a given year. In large ecosystems, even keeping track of the obvious is expensive, prone to error, and often done poorly if at all.

The concept of *indicator species* is one biological response to the problem of ecosystem complexity. Instead of trying to monitor all the dimensions of an ecological whole, focus on specific organisms that are sensitive to an important cross-section of those dimensions. Animals that depend upon a large number of other species are often selected for this reason: predators such as the spotted owl, whose well-being depends upon a large and diversified food web, are useful indicators of the health of the whole food web—and thus of the ecosystem that the food web inhabits.

Indicator species are messengers of well-being or ill health. Like all messengers, they risk being misunderstood—that is, confounded with the message itself. The bitter dispute over the spotted owl in the Pacific Northwest is really a battle over land use—whether the old-growth forest remaining in the Northwest should be harvested over the next several decades, or preserved for biological and recreational purposes. But it is disguised as a battle over whether the spotted owl, which depends upon old-growth forest habitat, is endangered. The real issue is not just the extinction of the spotted owl, but the extinction of its habitat.

Conversely, the popularity of salmon in the Columbia basin may or may not reflect a concern for the habitat of the wild fish. If hatcheries can feed commercial, tribal, and recreational fisheries, the species will no longer be an indicator of ecosystem health. Techno-logical advances in fish culture can change the environmental stakes.

Experimenting without a Laboratory

Governing large ecosystems is a matter of learning without teachers and experimenting without a laboratory. That does not mean there are no guideposts; the experimental method has been around long enough for its fruits to have changed the conditions of human life. But the guideposts do not have distances marked on them: what may be easy in a laboratory may be impossible at the ecosystem or population scale, and vice versa.

My purpose here is not to draw up a primer on experimental

ecology, but rather to sketch a policy-level discussion of the value, costs, and vulnerabilities of adaptive management *even* in an ideal setting. At the end of the chapter I return to the nonideal reality in which learning must take place.

Why Do It at All?

Ecosystem-scale experimentation is obviously difficult. Why—and, more specifically, when—is it worthwhile? There are three circumstances in which large-scale experimentation is worth contemplating. First, large ecosystems have some properties that cannot be observed at small scale. The abundance of salmon in the Columbia is a *different* aspect of the ecosystem from the abundance of salmon returning to a particular stream or hatchery. The $130 million being spent every year to rehabilitate the Columbia's salmon runs is not targeted at supporting any particular population, but at the entire population of salmon; indeed, the water budget, the largest single component of the annual cost, cannot be effectively allocated to specific stocks—if it benefits any, it benefits all that migrate while the river's flow is augmented.

More broadly, general-equilibrium effects appear only at the aggregate level. In economics, a rise in price may or may not be met by a decline in demand over time. A short-term observation can see how people cut back, but it will often miss the process of developing substitutes. In an ecosystem, the abundance of species may indicate little about the stability of the community of which it is part: diverting salmon production away from natural streams and into hatcheries, for example, has drastically cut the food supply of bald eagles, which once fed on the carcasses of spawned-out salmon. The salmon may be doing all right, but the ecosystem may become poorer as the American national symbol finds it harder to feed its young in summer. Effects of this kind cannot be stumbled upon in the laboratory.

A second reason for large-scale experimentation is that some effects, though visible in principle, are too small to observe at the laboratory scale. This is one way of understanding why industrialized countries have paid so little attention to the release of new chemicals into the environment. Unless an immediate toxic effect such as the death of test animals is observed, it may be difficult or impossible to detect an adverse effect until the "experiment" (in this case a trial without deliberately designed controls, so that the results are much harder to assess) is run on the population at large. Some fisheries scientists believe that hatcheries may be rearing "dumb" fish, which grow up without learning to dodge predators or to hunt effectively for food. Comparing their survival rates with those of broodmates brought up in the school of hard knocks by being released at an earlier life stage can test this conjecture—but it requires using the ecosystem as a sorting mechanism to see which population does better.

Third, ecosystem-level interventions may be present already because of policy decisions, or policy may be unable or unwilling to wait for the acquisition of information in the laboratory when it can be acquired more quickly by ecosystem-level change. When the purposes of human societies dictate action, ecosystem-scale experimentation may be the appropriate tool even without the favorable conditions found in the Columbia basin. For example, recognizing that a species is endangered can motivate quick action that might be too expensive to contemplate if extinction were not apparently imminent.

Internal versus External Validity

An experiment is a systematic way of answering a question. Whether the results constitute a valid answer is a test of the competence of the experimenter, although standards vary across fields. Most measurements in physics are done to accuracies of better than 1 percent, but the precision of opinion polling is rarely better than 3 percent. Greater precision requires a large (and costly) sample and may reflect poorly on the professional acumen of the pollster.

Adaptive management is field science; its laboratory is not a controlled setting, but a noisy, changing world of human actions and natural fluctuations. In a well-known essay on field studies in the social sciences the psychologist Donald Campbell developed a helpful distinction: between threats to *internal* and to *external* validity. Campbell noted that interpretations of field observations can be criticized on two quite different grounds. The criticism "You're wrong"—*something else really caused the observed effect*—is a threat to internal validity. The criticism "So what?"—the explanation is all right, but the *interpretation is not relevant* to the instances that count—is a threat to external validity. These potential challenges are the stuff of any debate over explanations, from the back porch to the scientific journals. Reliable knowledge rests on correct inferences (internal validity) correctly applied to other situations (external validity).

Threats to internal validity. The first three threats listed in Table 3–3 are familiar. In the field, control of the experimental setting is always imperfect, so something other than the experimenter's intervention could have caused the result observed (history). One aspect of the experiment that can be problematic is what would have happened anyway (maturation), and what might have happened by chance (instability). The question of what might have happened by chance is the only one of the threats that can be tested by statistics, the usual safeguard employed in science. The other entries in the table are less familiar.[2]

TABLE 3–3. Internal threats to experimental validity (after Campbell 1969).

History: events other than the experimental intervention are the real cause.

Maturation: the effect would have occurred anyway as a result of forces already in progress, such as growth or fatigue.

Instability: "it was a fluke" due to unreliable measuring tools or fluctuations.

"Hawthorne" effects: the effect is inadvertently caused by the experimenter.

Instrumentation: "fooled by your tools"—an unreliable measuring instrument may produce changes in the measurements obtained, when there is actually no effect.

Regression artifacts: apparent shifts occur when subjects have been selected on the basis of their extreme conditions.

Selection: biases result from recruitment of experimental subjects or controls.

The term *Hawthorne effect* originated in a famous study of industrial workers in the late 1920s, at a Western Electric Company plant in Hawthorne, Illinois. While carrying out carefully designed experiments to improve productivity by changing the work environment, a team of social scientists discovered that *any* change in work setting—including changes that went in opposite directions such as a rise in lighting level in one room and a decrease in another—produced a temporary increase in productivity. The observed effect (higher output) had nothing to do with the experimenter's intent (work conditions leading to increased output). Upon asking the workers why this had happened, the scientists discovered that the work crew were so pleased at having been singled out for special attention that they made a special effort to do their best. A Hawthorne effect may at first seem unlikely in the natural environment, but, as every frustrated bird-watcher knows, the very act of making observations in the wild can disrupt the natural behavior one is trying to observe. Similarly, the reliability of "instrumentation" needs to be explicitly considered whenever human judgments are part of the estimation procedure, as they usually are in environmental science.

The problem of regression artifacts is both stubborn and subtle. An example helps to describe the phenomenon: children of tall mothers tend to be not as tall; children of short fathers tend to be not as short. There are many factors that control height, of which inheritance is one. But the other factors tend to be random in their effect, some increasing and others decreasing height in any given child. So the shift is toward the average, or mean. This tendency, called *regression toward the mean*, is a well-known aspect of the statistical behavior of any population. Its importance in everyday life is that people usually do not form their beliefs on the basis of statistics. Rather, we impute causes to changes we observe, whether or not there is a sound reason to do so. The alternative of saying that the effect had *no* cause, but was simply the result of statistical fluctuation, is psychologically difficult to accept, particularly in something as deliberate as an intentional intervention. But it happens all the time: social programs *usually* show no effect that can be supported in statistical analysis. This is one reason statistical tests are a good idea—to provide a way of testing the chance that an alleged "effect" is a phantom. In general, however, threats to validity cannot be met by statistical checks alone; qualitative questions of design and interpretation are not matters that can be solved by crunching the right numbers.

The significance of regression artifacts in environmental science may seem less important than in the allegedly "softer" social sciences, until we recall that environmental remedies are applied precisely because some aspect of the environment is in an extreme state. Regression to the mean predicts, for example, that after a species has been declared endangered it will tend to become more abundant. This is not an effect at all, but a reflection of the fact that the human decision to declare a population in bad trouble is based upon its being *in extremis*. To the extent that that condition is caused by a variety of factors—as is virtually always the case in the natural setting—some of them will fluctuate in the next year, and the fluctuations will on average tend to bring the population up. In the early years of the Columbia basin program, before any of the rehabilitation measures could be carried out, there was a resurgence of salmon populations from the historic lows of the late 1970s, when the Northwest Power Act was passed. It took a special effort of political will *not* to take credit for this change, even though there was as yet no cause to which such an effect could be attributed.

Selection biases are subtle as well. The phenomenon is plain to see in the abstract: a real effect will work in *any* population, so its true strength will be seen in a *randomly chosen* population. But most populations in a policy setting are not random. Either they are selected by themselves or the experimenter, or they are drawn from a population restricted in ways that cloud the interpretation. Environmental rehabilitation is not sought at random, but to remedy specific problems or losses. Experiments done on the salmon that have survived in the Columbia drainage may yield results that are misleading because they are not representative of the larger, rebuilt population being aimed for. The difficulty is twofold. First, it is impossible to avoid some selection biases: the only salmon in the Columbia are those that have survived dams, fishing, and habitat loss. Second, even when it is possible in principle, randomization can be ethically and politically difficult, whenever it involves denying someone a supposedly beneficial treatment for no reason other than experimental protocol.[3]

Threats to external validity. The "so what?" challenges (see Table 3–4) involve allegations of two kinds: that the design or implementation of the experiment made the results misleading or inapplicable elsewhere; or that the experimenter was doing something different from what she thought she was doing, so that the effects are not in fact due to the causes to which they are attributed but to something else.

These threats to validity must be taken seriously in virtually all cases of adaptive management. The logical response to doubts about

TABLE 3–4. External threats to experimental validity (after Campbell 1969).

Measuring affects the subjects in such a way that subsequent measurements can no longer be generalized.

Selection of subjects means that response to experimental treatment is no longer representative of total population.

Where multiple interventions are jointly applied, the effects may not indicate the response to separate interventions separately applied.

The measurements are complex and include irrelevant components that may produce apparent effects.

The interventions are complex, and replications of them may fail to include those components actually responsible for the effects.

external validity is to try the treatment in another place, since this directly puts to a test the question of whether cause and effect can be reproduced somewhere else. This may often be easier to do in an environmental setting than in the arena of social policy, where reproducing a treatment often depends upon enactment of legislation or other social processes that are impossible for an experimenter to pursue predictably.

The most important aspect of the threats to validity is, however, that they are threats, not fatal wounds. Campbell ended his discussion of validity this way: "This is evaluation, not rejection, for it often turns out that for a specific design in a specific setting the threat is implausible, or that there are supplementary data that can help rule it out even where randomization is impossible. The general ethic, here advocated for public administrators as well as social scientists, is to use the very best method possible, aiming at "true experiments" with random control groups. But where randomized treatments are not possible, a self-critical use of quasi-experimental designs is advocated. We must do the best we can with what is available to us." Insisting on an idealistic approach to science does not entail refusing to do science unless it is invulnerable to criticism. It entails approaching problems scientifically, so that reliable knowledge can be obtained and the frailty of knowledge assessed; deciding not to experiment is a choice too. The concept of validity provides an orderly framework in which to make judgments of reliability.

Dependence on Statistics

Validity is a qualitative basis for analyzing experimental design. Usually there is a quantitative dimension too, for which statistics is used as a framework to judge the quantitative reliability and credibility of measurements and inferences. Statistics is an area that policy-level discussions habitually avoid—on the theory, perhaps, that numbers are tractable enough to be taken care of by the technicians. But although the complexity of statistical analysis makes the technicians necessary, their ministrations are not sufficient when the control and execution of experiments entail cooperation among institutions. That is the case in adaptive management. For some policy questions, statistical concepts promote understanding of the nature of the policy judgments required.

Type I and II errors. The scientist is an idealist, someone who does not want to claim that something is true that turns out later to be false. This idealism is characteristic of science as a human endeavor. It is what makes astronomy basically, instantly different from astrology. It is what caused Newton to wait two decades to publish his masterwork, the *Principia Mathematica*; he had to reconcile his estimates of the orbit of the moon with the data. The reluctance to claim that something is true unless hard proof is at hand is not restricted to science. The Anglo-Saxon notion that a person should be presumed innocent follows a similar rule: someone accused of a serious crime must be shown to be guilty beyond a reasonable doubt. The underlying principle is that it is better to let a guilty person walk free than to convict one who is blameless.

Science and criminal law are designed to avoid Type I errors; a definite conclusion (a scientific proposition or guilt) is accepted only when there are good grounds. A Type I error is an error of commission—affirming a proposition that turns out to be false. Avoiding such errors entails tolerating errors of omission: it is acceptable for some guilty persons to go free; it is acceptable for some scientists' theories to languish unpublished even though they may be true. In an unchanging world, the accumulation of reliable knowledge depends upon a bias against Type I errors.

The problem is, we live in a changing world. It is a world in which forces already in play will bring about unwelcome results unless they are channeled or regulated. Consider new chemicals, tens of thousands of which are synthesized each year. Should we permit their release into the environment unless toxicity is shown? That would parallel the rule of "presumed innocent": presumed harmless until shown to be toxic. Consider endangered species. Should we wait to intervene until we have a sure method of saving the population that is left? What if it perishes while we are doing research?

In these cases, there is a cost to *not* acting. Failing to act when it turns out to have been necessary is a Type II error: to reject as false something that later turns out to be true. Such a rejection does not pose a problem for society when a single scientist's bruised ego is at stake. It can pose a problem for victims of crimes who see their malefactors walk free, perhaps to hurt the victims again. And it can be a problem for all of us, when an endangered species goes extinct because a sure way of saving it did not exist, or when a toxin is released because there was no good reason to control it.

Are there circumstances in which Type I errors should be avoided, and others in which Type II errors are a problem? Their differences of logic suggest that it might be possible to sort out situations, and to design institutions to apply the appropriate burden of proof. This happens, to a degree. One does not apply the standards of criminal law—under which the state accuses individuals—to civil disputes, where neither plaintiff nor defendant is categorically entitled to presumptions.

But two problems remain. The first is the problem of boundaries. As the plight of the crime victim demonstrates, the presumption of innocence has consequences for parties other than prosecutor and defendant. There has been bitter dispute over capturing endangered California condors, whose numbers were reduced to fewer than two dozen, in order to breed them in captivity. The problem is whether to risk a Type II error—allowing the condor to go extinct when it could have been saved—or to chance a Type I error—killing the last condors in an attempt to preserve their kind, when they might have survived unaided.

A second problem may be more tractable: the misunderstanding of the nature of the policy problem. Scientists, with their aversion to Type I error, have traditionally trained policy analysts and decision makers to think in terms of avoiding Type I errors. But the policy may not be one in which Type I errors matter most. In drug regulation, for example, those suffering from terminal diseases are often denied access to medicines under development. Here, the responsibility of scientifically trained regulators to protect the patient population over the long run (by not declaring safe something that turns out to be dangerous) comes into conflict with the principle of affording the doomed a chance to try something that may be a longshot but is at any rate an alternative to certain death. There will inevitably be problems in defining who qualifies as terminally ill under a regime that treats them differently in providing access to medicine, but those problems may be more malleable than the ones that now trouble regulation.

In the environmental arena, fisheries scientist Randall Peterman has clarified the implications of Type I and Type II errors, illustrating his explanations with cases in which scientifically trained specialists delayed harvest reductions until large declines had occurred—because they were reluctant to claim that something was true (the population was decreasing) until their evidence was solid. In one case, the harvest involved whales, and rules that allowed high levels of whaling until damage was clearly established were adopted in 1983 by the International Whaling Commission. This is not an obscure academic question.

Power of test. The cure for this problem is not fewer statistics but more: it is to make use of a statistical concept called "power of test." Power of test asks a penetrating question: When the evidence is weak, how large an effect might be missed, given the data collected? If the population of whales fluctuates a lot, and a change has been seen, how large a decline *might this turn out to be* given that the fluctuations can hide a lot of change? More generally, what are the relative costs of a Type I error and a Type II error, given the data available? Ordinary science asserts that the cost of a Type I error is always very high. Firefighters respond to false alarms because, in their world, the cost of a Type II error is always very high. The power of test provides a quantitative framework within which to examine cases that are not so clear-cut.[4]

Persistent Challenges

Experimentation is difficult enough that some of our brightest students spend years mastering experimental methods in laboratories and the field. What are the most persistent challenges presented by the art of experimentation to adaptive management?

Risk of failure. The first challenge is to do experiments at all. Projects that manipulate large parts of ecosystems are ones whose failure carries real costs, particularly to those in charge. This is the habitat of two organizational species described by Campbell: "*Trapped administrators* have so committed themselves in advance to the efficacy of the reform that they cannot afford honest evaluation. For them, favorably biased analyses are recommended, including capitalizing on regression, grateful testimonials, and confounding selection and treatment. *Experimental administrators* have justified the reform on the basis of the importance of the problem, not the certainty of their answer, and are committed to going on to other potential solutions if the one first tried fails. They are therefore not threatened by a hard-headed analysis of the reform." It is noteworthy that the administrator may be trapped by his or her situation, even if the administrator wants to act in an experimental mode.

Of course, even trapped administrators cannot guarantee the success of their projects. As a result trapped administrators may also be interested in experimentation—for the purpose of making themselves look good. There is accordingly a moral hazard for adaptive management: that managers will cook the books. The possibility is thinly veiled in Campbell's advice to trapped administrators: skewed science can be beneficial to the trapped administrator, giving the appearance of rigorous evaluation and testing but providing a predetermined positive result.

The problem is unlikely to arise through outright corruption, but rather in the mix of experiments to be tried. Since resources are perennially short, one can easily choose a set of experiments, some of which are meant to test the effectiveness of environmental rehabilitation rigorously, and others designed with a few methodological corners cut so as to present management with a defensible track record. Corners must be cut anyway, and some aspects of the overall strategy are more dubious than others. This line of thinking is attractive not only to the administrator—who may not know much about science and experimental design—but also to the scientist interested in adaptive management, sincere about rehabilitating a large ecosystem, and pragmatic in his or her appreciation of the political situation of the administrator, who is a necessary ally. Corruption will not come from outside the circle of science, but from within, and it will be clothed in the plausible vestments of technical judgment and pragmatism. This is a persistent problem, inherent in the task of adaptive learning in large ecosystems. It can be recognized and managed; it cannot be banished.

Randomization, selection, and controls. The search for reliable knowledge is often compromised at the outset by the need to apply limited resources to a landscape that is more in need of help at some places than at others. In a laboratory the experimenter can randomly choose subjects and controls, but in adaptive management, each subject or site is in a different place, and control sites—the ones deliberately left alone—are deprived of a project that is presumed to be beneficial. Since ecosystems are both complex and unique, identifying clear-cut control cases is necessarily a matter of judgment. Yet controls are still essential: experiments without controls face severe threats to both internal and external validity.

Scarcity can help the experimenter by making a virtue of necessity. If a desirable intervention, such as a new hatchery, is too expensive for as many to be provided as are wanted, sites that are waiting can serve as controls—but only if the sequencing of sites is based on awareness of selection effects, and, most important, if the control sites are monitored to see what changes there in the absence of intervention.[5]

Delay and cost. Even after adaptive management is adopted as policy, it must remain in place over times of biological significance if it is to yield reliable knowledge. Such an approach calls for patience, which can be hard to sustain when there are political suspicions about delays and costs. Experimentation is a form of study, and study is a form of delay. Critics of the Columbia basin program have charged that adaptive management is a cover for delaying the rescue of endangered environments. The fact that those using an adaptive approach necessarily cooperate with those who manage large technological systems tends to raise suspicion more; the political pragmatism of the adaptive approach invites skepticism from critics. But from another point of view, adaptive management legitimates actions that are explicitly *un*certain, because of the value of the knowledge they would produce. Questions about delay should logically be framed in a comparative fashion: Is it slower than practical alternatives? Given the large scale of ecosystem interventions, the delays arising from a carefully planned experimental protocol turn out to be minimal.

In any event, the accumulation of reliable knowledge often feeds itself, as growing understanding leads to faster progress and more rapid learning. In economics a similar idea is similarly hard to sell—saving and investment, though hard at first, lead to riches faster. In fact the start has been painfully slow in the Columbia basin—ironically, because the awareness of uncertainty and demand for knowledge have risen quickly. The experimental hatchery in the Yakima Valley of Washington state was initially proposed in 1981. Each time the project has been brought before the planning council, the requirements for learning have been raised—at my own behest in 1984, when I argued, successfully, that the hatchery should become an experimental facility. More recently, concern over the genetics of hatchery fish has led to further elaboration of hypotheses that need to be tested, resulting in more redesign. "This is like opening Pandora's box," council fisheries biologist Harry Wagner said not long ago. "There appears to be an endless supply of difficult questions inside."

Cost is another problem. The information needs of an experimental approach are high. The most desirable experiment is one that tests as few things as possible, so that the relationship between cause and effect can be isolated as clearly as natural fluctuations will allow. That is the point of a laboratory: to provide a controlled setting in which as few things vary as possible, and those few are under the control of the experimenter. In the field, knowledge and monitoring substitute for control. But knowledge must be gathered and analyzed so that it is usable. The kinds of knowledge appropriate to ecosystem management tend not to have been gathered in the past, because people were exploiting the ecosystem rather than trying to manage it. So the adaptive approach must make a large down payment to start with, to amass enough of a baseline to comprehend the outdoor "laboratory" in which the experiments will be launched. This process can be slow and expensive. The Columbia basin's system planning process took three years and several million dollars, simply to assemble enough information to make decisions.

Whether these challenges will prove to be insuperable is a question that varies with time, institutional arrangements, and the determination of the experimenters. They are not insurmountable in principle. The progress of science itself demonstrates that. But in practice there are few places where the combination of experiment and management described here can be said to have gained even an initial foothold.

Adaptive Management in Real Human Institutions

Experimenting at the ecosystem level is big science: the "laboratory" of the field setting is usually large in scale, the interventions are expensive, and they require large organizations to deploy and operate. The discussion of experimental design so far has assumed that the experimenter is a rational individual and is supported by a rational, responsive set of institutions, ready to carry out the experimenter's instructions on how to manipulate the ecosystem. These assumptions were vital to understanding the challenge of doing ecosystem-scale experiments.

Yet those assumptions are unrealistic. The experimenter in adaptive management is not a lone scientist, but a member of a management structure. Usually the experimenter will be a team, not an individual, with the internal questions of coordination faced by any team. And the institutions that constitute the "apparatus" for the experiment are organizations with multiple goals, of which social learning is only one and usually not the most important one. Four limitations on institutional learning and responsiveness are of key importance: reliance on operating agency staff, the disruptive capability of policy changes, vulnerability to political change, and the requirement that the adaptive manager be an able negotiator as much as a visionary scientist.

Institutional Vulnerabilities

Regular program staff. For a policy to be an experiment, regular program staff must carry out major portions of the experimental protocol.[6] Two problems arise: they may not know what they are doing and bungle the task, or they may know and subvert it instead.

First, the experimenter faces the problem of training research assistants who already have full-time jobs and who are likely to see adaptive experimentation as having to do more—more record-keeping at a minimum—and having to change established routines. Experimentation entails following the instructions of research people who are not part of the regular chain of command. But if something goes wrong, it will be the regular line organization that takes the blame. Accordingly, experimenters need to have the confidence of the regular line organization, and they must keep on earning that confidence by respecting the imperatives that impel discipline in those institutions. That confidence is a precondition for training staff in their research tasks.

In an adaptive situation, the lessons learned from experiments are likely to affect the careers of the regular program staff. They therefore have a stake in the outcome that can skew their performance of research tasks. These biases depend upon the staff members' perceptions, not the experimenters'. So the experimenter needs to understand the perceptions of the "research assistants" and take steps to guard against bias. The bias can take the form of efforts at helping, as at the Hawthorne plant, as well as sabotage or resistance. But as with training, the key point is to realize that human beings living and working in an operational environment, who do not see themselves as researchers or guinea pigs, are being asked to be one or both. Their responses cannot be assumed to be simple, and their willingness to communicate with the experimenters is vital.

The experimenters' relationship with the operational staff may be most critical *after* the experimental phase. If a change is to be made as a result of lessons learned from the experiment, the acceptability of that change to those who must implement it is likely to be heavily influenced by their memory of the experiment and how it was carried out.

Reliance on the regular staff also has a feedback effect, limiting the range of experimentation that can occur. The reluctance of implementing organizations to carry out experiments that are feasible is a problem especially with respect to power of test. Large changes are necessary to see *anything* above the interference of large natural fluctuations, but from the operators' perspective large changes are also likely to endanger the primary mission of the agency and to conflict with the way things are usually done. The water budget represented a large change for hydro project operators accustomed to maximizing power revenues; for a decade after the water budget was invented, there was active resistance each spring to carrying it out. The larger pattern follows: experimenters with plans for large changes will tend to lose the confidence of the operating staff; conversely, experimenters who want to work at the ecosystem scale will tend to trim their plans in advance. This is the feedback effect— the most readily implemented experiments tend to be the least powerful probes of the ecosystem.

A corollary is that finding creative ways of obtaining powerful tests without forcing operating staffs to do things they think are wrong or foolish is of central importance to the human part of experimental design. Taking advantage of changes being forced by nonexperimental considerations can be useful. The Columbia basin program's focus on threatened and endangered salmon stocks has already produced strenuous attempts to release more water in the semiarid Snake River basin. There is conflict over these reallocations, but also a significant opportunity to carry out experimental programs of water releases that could probe the economically important question of when higher flows actually produce biological benefits. This is a circumstance in which operational staff are already bruised by changing conditions, and may welcome experimenters who approach them with an understanding of their confusion at having their jobs altered and criticized.

Turbulence. The example of changes in Snake River flows highlights a second problem of the ecosystem laboratory: changes caused by human intervention. In contrast to natural fluctuations such as droughts or insect infestations, changes resulting from policy are unlikely to be random in a statistical sense. So to the unpredictability of a natural system subject to fluctuations one must add the possibility of policy changes that in effect tilt the apparatus or change its settings without the foreknowledge of the experimenter.

As with threats to validity, the issue is how to judge rather than when to reject. Some policy changes can be taken into account smoothly because they alter the values of identifiable variables (such as water flow) in simple ways. These changes often enhance power of test by making the observed range of inputs larger. But other policy changes trigger pervasive changes in human use and behavior that can threaten validity on several dimensions at once. The shifts caused by the assertion of Indian treaty rights, for example, changed the economy of salmon fishing throughout the Pacific Northwest and Alaska. The widespread changes in harvest practices, many of which are not documented, would have made it difficult to study harvest pressure on salmon stocks in an experimental setting during the period when these changes were diffusing through the fishing industry.

Turbulence in human management is likely to be a permanent feature of ecosystems heading for sustainability. The same forces that make an adaptive approach compelling—uncertainty about the responses of both the natural and social systems—also make reactions other than experimentation politically attractive. Developing the skill of designing, redesigning, and operating large experiments under these conditions is a challenge already facing some oceanographers, conservation biologists, and atmospheric scientists who are studying phenomena affected by human actions and policies. It is a kind of skill that university science, with its laboratory orientation, has as yet done little to develop or even to recognize; here, adaptive management is likely to be on the cutting edge of scientific method, with the risks and excitement implicit in that situation.[7]

Political vulnerability. Adaptive management not only needs to respond to and anticipate turbulence; it can become a focus of turbulence. Randall Peterman has described research on British Columbia sockeye salmon in the 1930s that led to the conclusion that hatcheries there did not work. Under budgetary pressure because of the Depression, the hatcheries were closed. A reanalysis of those data 50 years later showed that the statistical power of the data collected originally did not warrant the conclusion that the hatcheries did *not* work—they showed only that the experiments were not powerful enough to prove that the hatcheries *did* work.

This is an object lesson about the importance of power of test, but it also illustrates the consequences of doing research. Research that has consequences is research that actors will try to tamper with or keep from occurring. Adaptive management is research that must have consequences if it is to be worth the high costs of doing it.

A nefarious motive need not be involved for turbulence to exist. The Northwest Power Planning Council raised sharp questions in

1989 about the rising cost of the Yakima hatchery authorized under the Columbia basin program. The high costs at Yakima were due largely to its experimental nature. Extra rearing ponds for juvenile fish and a more expensive plumbing system were required, so that populations could be kept separate from one another in case of disease or if water temperature needed to be varied. Monitoring costs were high. The council approved the Yakima budget, after grumbling about whether the research was worthwhile. "The depth of the region's commitment to learning is unclear," an informed insider wrote in 1992. When a body with an explicit commitment to adaptive management and unusual budgetary discretion expresses such reservations about its leading experimental facility, one is forced to consider the magnitude of the problems other large-scale experiments are likely to have elsewhere.

Flexibility and negotiation. The institutional conditions of adaptive management recall a lesson learned at the Hood River conservation project: flexibility and negotiation are of instrumental importance. These issues are taken up more systematically in the next chapter. Here, I note that the experimenter in adaptive management is not a lone scientist laboring in obscurity. He or she is much more the late–twentieth century scientist-bureaucrat-adventurer, the kind of figure who leads easily in large organizations, to whom budgets are as much a part of the intellectual furniture as is Darwin's theory of evolution, whose ambition is expressed more through creative diplomacy than through abrasive brilliance. This may be a surprising and perhaps disappointing picture for those who think of environmentalism as Thoreau-like contemplations. But since Aldo Leopold founded the first professorship of wildlife management at the University of Wisconsin in the 1930s, the link between organizational success and environmental science has become stronger as the magnitude of the problems to be encountered has more clearly expanded to the scale of industrialism itself.

Table 3–5 is an annotated version of Table 3–2, the list of conditions favoring adaptive management. The italics following each statement summarize the social dynamics and institutional rigidities that complicate an experimental approach, even under favorable conditions. These complications constitute a list of concerns to be addressed in the chapters ahead.

TABLE 3-5. Institutional conditions affecting adaptive management.

There is a mandate to take action in the face of uncertainty. *But experimentation and learning are at most secondary objectives in large ecosystems. Experimentation that conflicts with primary objectives will often be pushed aside or not proposed.*

Decision makers are aware that they are experimenting anyway. *But experimentation is an open admission that there may be no positive return. More generally, specifying hypotheses to be tested raises the risk of perceived failure.*

Decision makers care about improving outcomes over biological time scales. *But the costs of monitoring, controls, and replication are substantial, and they will appear especially high at the outset when compared with the costs of unmonitored trial and error. Individual decision makers rarely stay in office over times of biological significance.*

Preservation of pristine environments is no longer an option, and human intervention cannot produce desired outcomes predictably. *And remedial action crosses jurisdictional boundaries and requires coordinated implementation over long periods.*

Resources are sufficient to measure ecosystem-scale behavior. *But data collection is vulnerable to external disruptions, such as budget cutbacks, changes in policy, and controversy. After changes in the leadership, decision makers may not be familiar with the purposes and value of an experimental approach.*

Theory, models, and field methods are available to estimate and infer ecosystem-scale behavior. *But interim results may create panic or a realization that the experimental design was faulty. More generally, experimental findings will suggest changes in policy; controversial changes have the potential to disrupt the experimental program.*

Hypotheses can be formulated. *And accumulating knowledge may shift perceptions of what is worth examining via large-scale experimentation. For this reason, both policy actors and experimenters must adjust the tradeoffs among experimental and other policy objectives during the implementation process.*

Organizational culture encourages learning from experience. *But the advocates of adaptive management are likely to be staff, who have professional incentives to appreciate a complex process and a career situation in which long-term learning can be beneficial. Where there is tension between staff and policy leadership, experimentation can become the focus of an internal struggle for control.*

There is sufficient stability to measure long-term outcomes; institutional patience is essential. *But stability is usually dependent on factors outside the control of experimenters and managers.*

This chapter has defined a task: to describe a learning strategy that does not assume that adaptive policies in large ecosystems will be designed and executed by rational actors in an ideal world. This involves a two-way adjustment: on one side, to suggest institutional designs and practices that can compensate for the inevitable weaknesses and unavoidable failings of real institutions; and on the other, to temper and frame expectations of what is attainable in an imperfect world. It is the only world we have.

Chapter 4

Gyroscope: Negotiation and Conflict

> The reason boundaries exist where they do is that
> they are tested periodically.
>
> —C. S. Holling, *Adaptive Environmental Assessment and Management*

Disputes have been a characteristic feature of environmental politics, for reasons that make them indispensable—and inconvenient—in the search for sustainable development. Environmentalism flowered first in free societies and has been a way to challenge tyrannical regimes. For good and ill, freedom spawns environmental conflict. Those who benefit from access to or use of a place seldom experience the total costs of their activity. Some costs are displaced across space, time, and culture, to be inflicted on others; the result is dispute. When decisions force clashing values into collision and crack the insulation of powerful institutions, conflict can be indispensable as an integrating mechanism; its disadvantage is that the outcome may result in serious losses to some or all sides, something none can control well.

Conflicts in ecosystems can easily bog down. There are typically many parties, not the two opposing sides for which courts, our normal means of processing conflict, are designed. Environmental questions are complex and often lack definite answers. And there may be no imminent or apparent crisis, even if irreversible changes are under way that will make future crises much less tractable. The first task of this chapter, accordingly, is to clarify how existing institutions affect conflict and learning in making policy. The basic theme is that political conflict can provide ways to recognize errors, complementing and reinforcing the self-conscious learning of adaptive management.

When conflicts bog down, there are two alternatives. One is to alter the institutions that make authoritative choices. The Northwest Power Act created the Northwest Power Planning Council, a new arena in which disputes could be conducted. But when societies are unable or unwilling to define new institutional arenas in a timely way, less formal means of conducting disputes are useful, to organize and facilitate negotiation among disputing parties. In theory governments decide matters. But because governmental powers are limited in practice, negotiation remains an essential means of resolving disputes. Useful innovations have been made over the past fifteen years in fostering negotiation in environmental conflicts in the United States, and these are discussed here.

Conflict can either enhance or prevent learning. Because learning requires that observations be made over times of biological significance and spatial scales that transcend property lines and political boundaries, conflict can thwart the learning necessary to reach sustainability. Yet conflict is also indispensable to defining, over time, a socially sustainable order, because it impels institutions toward such a search in the first place.

Environmental Disputes

Origins

How people use the natural world and its resources has bred dispute for as long as there has been competition for desirable lands or waterways. But even if all land were equally bountiful, conflict would continue for a multitude of reasons. First, many natural resources will remain common property; a fish is not the property of the fisherman until caught. As a result fishermen try to catch as many fish as they can, even when they know that a dead fish can spawn no young for future seasons. The problem in these cases is not that individuals are being irrational. Quite the contrary: it is that they are being rational. In this situation what needs to be altered is the relationship among people and natural resources, so that what becomes rational for an individual is also sensible for the natural system and the human community. Second, the property rights that do exist are based on precedent. In places where precedent is based on use, landowners use their water to the maximum every year, knowing that any diminution can impair their claims in future years. Waste is institutionalized, and conflict made endemic. Third, using resources produces spatial spillover effects: one user's watering his or her land decreases the amount left in the stream for the next user downstream; industrial development brings wages to the factory worker but pollution to the plant's neighbors. Fourth, there are temporal spillovers; cutting down a forest precludes many options for a long time. Fifth, there are disagreements over how costs or benefits in the future should be compared with costs or benefits in the present, for example for the purpose of paying damages. This is not an exhaustive catalogue, but it illustrates both the durability of disputes over natural resources and the depth of change needed before we can rely on sustainable use of resources.

Most environmental disputes are opportunities for political mobilization; by definition, the environment tends to lie outside the control of individuals. Usually it is difficult or impossible for individuals to prevent or to deal with environmental damage: individuals can drive or not, but no one person can clean up polluted air. These are inherently political issues, and when governments or large private organizations have been reluctant to find collective cures to environmental ills, nongovernmental organizations have risen to compel action through political conflict.

Limited Government

Governments play a major role in environmental disputes for at least two reasons: because the scientific complexities of environmental problems can be illuminated or resolved only through publicly supported research, and because environmental quality is a public good: the benefits are distributed in such ways that it is difficult for naturally occurring markets to charge for them. Because there is often no way to pay, less environmental preservation takes place than people are willing to support. Conversely, both rationales and pressures exist for government intervention.

During the 1970s laws authorizing environmental protection were enacted rapidly at both federal and state levels. Legislation was complemented by judicial action as environmental law became a major new field of litigation and administrative procedure. In many cases, however, the role of governments remained anomalous. The traditional view—that the actions of a duly constituted governmental entity are by definition "in the public interest"—is rendered untenable when one or more governmental bodies is a disputing party. Local governments and functional agencies such as the Army Corps of Engineers, which operates the federal dams on the Columbia, are especially likely to be in this position. On the other hand, governmental agencies with the legal authority to make the choices at stake may find choosing hazardous: when contending parties in polarized conflicts make all alternatives politically punishing, especially for elected officials, a governmental agency can be paralyzed by conflict, unable to make a choice, not because it lacks formal authority, but because no available alternative is preferable to continued delay.

Environmental policy is therefore never far removed from politics in general. Unlike such policy arenas as basic science or the judiciary below the Supreme Court level, environmental controversy—despite its formidable technical complexity—has been covered by the mass media, is often a factor in elections, and has fueled an array of interest groups.

Environmental conflict had an ambivalent aura at the outset, which it has never lost. Like all conflict, environmental disputes challenge the social order. As those attacked for polluting the environment point out, environmentalists strike at the heart of the economy, at growth and jobs. But environmentalists have maintained their standing as well as any critics of the establishment, claiming to represent an authentic struggle for justice. "Those who profess to favor freedom and yet deprecate agitation are men who want crops without plowing up the ground. They want rain without thunder and lightning. . . . Power concedes nothing without a demand. It never did and never will," wrote the slave leader Frederick Douglass; environmental activists have often brought the rain.

Conflict and Learning

It is a different question, however, to ask whether environmentalism has brought learning. Learning and conflict stand in a contingent relationship; here I survey several theories of policymaking in an attempt to estimate the dimensions of the contingencies. Theories of public policy are a sensible focus for this inquiry because if social learning is to occur in a durable way it must be reflected in policy. These theories show that learning can occur via political conflict, although its salutary effects must be considered in light of the hazards of conflict.

Experts as Teachers

A useful baseline is provided by the American philosopher John Dewey, one of the leading intellectual figures of his time, whose 1926 lectures at Kenyon College appeared the following year as *The Public and Its Problems*.[1] Dewey's analysis remains strikingly pertinent today, although the precision of his language and his optimism no longer seem attainable.

The issue Dewey tackled was democratic governance in an industrial society. Like most who have written on this subject, he focused on the puzzle of how a mass society could deal with questions that necessarily required expert knowledge to articulate or decide. The answer, he argued, lay in understanding the contributions that citizen and expert should make to the formulation of policy. Dewey observed that the importance of the individual in the political realm had grown enormously during the nineteenth and twentieth centuries, the period in which large organizations in business and government acquired unprecedented power. Paradoxically, the citizen's political influence was undercut just as it became central. The mass societies of industrialism made everyday living radically interdependent: no one finds his own food, builds her own shelter, or provides his own transportation in the modern world; these and other necessities are distributed via the mass markets of industrialism, as products and services that are produced and sold by large, impersonal institutions. The social character of industrialism, Dewey concluded, could not foster a knowledgeable citizenry with a strong sense of political community. Individual persons literally could not understand the details of how their daily needs were met; therefore, they could not know their own interests. Faced with a reality of such intimidating complexity, the natural tendency was to cling to doctrines that provided reassurance, even when they did not describe the world accurately. In our own century we have seen in both communism and anticommunism the simplistic faith that Dewey pointed to, undermining democracy.

In place of a public life devoted to the battle of dogmas, Dewey called for an experimental politics. "Such a logic involves the following factors: First, that those concepts, general principles, theories and dialectical developments which are indispensable to any systematic knowledge be shaped and tested as tools of inquiry. Secondly, that policies and proposals for social action be treated as working hypotheses." That sort of policy would be one in which democratic processes would "uncover social needs and troubles"; the distinctive strength of citizens, he suggested, lay in their ability to recognize problems. Then, with an agenda set by democracy, "The essential need is the improvement of the methods and conditions of debate, discussion and persuasion. That is *the* problem of the public." Debate would require the participation of experts, but they would act in a special way: not to render judgments, but to analyze. If experts *acting as teachers and interpreters* could decipher the technological world for citizens and enable them to make sensible political judgments, the constitutional mechanisms that are designed to advance the public interest over selfish interests could work as originally designed.

Social research since the 1920s has made Dewey's confidence about the rehabilitation of democracy seem less plausible. Three problems affect social learning. First, the citizenry seems to lack the capacity to make many of the judgments Dewey called for. He argued, for example, that voters should judge their rulers by the consequences they produced. Today it remains questionable whether, even with expert assistance, voters can perceive or evaluate consequences that are distant (such as foreign policy) or delayed (such as budget deficits).

Second, like others whom we identify with the Progressive movement at the turn of the century, Dewey assumed that experts were disinterested, objective specialists. Today, when scientists and artists have been accused of fraud or evil before congressional committees, we are aware of and cynical about the tensions that cloud the relationship between the public and any expert.

Third, public attention has displayed little durability and stability in the face of subtle and complicated scientific problems, a long public agenda, and intense competition to attend to many other things in life besides affairs of state.

Each of these weaknesses has an answer from contemporary political science, as I argue below. Those answers are important because Dewey's analysis defined governance in two senses that are important for our purposes: he emphasized that learning is of central importance, and he showed how the wider topic of democracy frames any discussion of policies in specific areas such as environmental management. Today's theories portray a world with little learning or democracy, in the sense in which Dewey understood those words. But with the aid of those theories, we may assess the feasibility of the kind of governance Dewey thought was essential.

Attention Span and Tragic Choice

On the possibility of holding government accountable for the consequences of its actions, consider Anthony Downs's 1972 essay, "Up and Down with Ecology." Downs, an economist with a flair for explaining political processes vividly, argued that environmentalism was a passing fad because it was an insoluble problem in the terms stated, and therefore the media and public would lose interest. His analysis is now part of the conventional wisdom as a criticism of the mass media, although his prediction that environmentalism would fade has yet to be confirmed.

What has proved more durable is the idea that the mass media enforce a limited public attention span. We know of the wider world beyond our personal lives only through the media, and, Downs pointed out, the media require viewers or readers as a condition of economic survival. This combination puts great weight on whatever is immediately salient. As a result, the world projected is one in which the sensational may equal or exceed any other dimension of worth. Problems of the environment were novel, vivid, and newsworthy in the months following Earth Day in 1970, but because they are rooted in economic activity they cannot be solved easily or quickly. Hence their news value would diminish, and with it public attention. In retrospect, what Downs missed was the scope of the word *environment* and the persistent immediacy of threats to health and beautiful places.

What he may not have missed is the way in which Dewey's call for political accountability based on consequences has been refocused in a society that sees reality through the mass media. While environmental concern has been buoyed by a steady stream of bad news from industrialism, individual environmental concerns have gone up and down as Downs predicted. The environmental era is now old enough to have had two separate bouts of concern about the greenhouse effect, and two waves of activity aimed at protecting the ozone layer in the upper atmosphere from gases released from spray cans and other industrial products.

Downs's emphasis on the brief political attention span forced by the media suggests that learning is all but impossible if it must take place at the level of the general citizenry. The issues are complex enough to thwart a division of the world into good guys and bad.[2] But the cycling of issues forestalls the building of more sophisticated conceptual frames.

A loophole Downs saw is crucial, however. If the brief concentration of mass attention on an issue produced institutional change in government, then even after the media searchlight passed on, there might be a lasting effect. Congress enacted the Endangered Species Act several months after Downs's article appeared; its history since then illustrates the lasting power of institutional arrangements.

The Endangered Species Act entails "tragic choice." This term, invented by legal scholar Guido Calabresi, identifies decisions that force a society to choose between its fundamental values. Consider selecting soldiers from among the young men of a democracy, or rationing lifesaving medical care, or deciding how to limit the size of a population: each presents life-and-death circumstances in which deeply held values clash. In such situations societies "must attempt to make allocations in ways that preserve the moral foundations of social collaboration." It is often possible to transform choices by allocation methods that displace or disguise their life-taking nature, as when government advertises for recruits by portraying military service as a form of occupational advancement. Yet such devices are only illusions, for the underlying conflict of fundamental values poses "a prospect of insuperable moral difficulty, a nightmare of justice in which the assertion of any right involves a further wrong, in which fate is set against fate in an intolerable necessary sequence of violence."

The Endangered Species Act declares that the loss of a genetically unique species is, "quite literally, incalculable." When such a loss is threatened, heroic means must be used if necessary—actions taken without regard to cost. In the Columbia basin four stocks of salmon have now been declared threatened or endangered. It is not yet clear how they will be protected, or at what cost; one cost is certain to be much more fractious relations among those who operate the river's dams, the utilities who want to produce as much power as possible, and the fisheries managers whose priorities are to return the river to an annual cycle as much like its damless past as possible. The act operates in a compelling fashion, by identifying a population being extirpated and forcing an explicit decision by officials of the federal government. The choice is framed so as to highlight the incalculable biological loss: the price of further economic exploitation is death for a species. Protection can be a difficult choice when other values are at stake, perhaps tragic for those like Indian harvesters whose communities could be shattered by a sudden suspension of fishing to protect a declining stock. Thus the pursuit of a right can entail a further wrong.

In this sense, the Endangered Species Act institutionalizes tragic choice. By declaring the preservation of species a fundamental value, the statute widens the range of resources society may be compelled to commit to that purpose. Where those resources suffice to save the listed species, a price may be paid, but a goal will have been reached. Where the marshaling of those resources imposes costs on identifiable persons, and their fundamental values are violated, a further wrong has been done. And when the resources marshaled fail to save the imperiled species, all have lost important ground.

Calabresi noted that attempts to cope with tragic choices through fixed policies can be inherently unstable. The instability illuminates a connection between conflict and learning. When the provisions of the Endangered Species Act are invoked, there is no time left to temporize; the fact that a species is at risk of extinction requires action without regard to cost, including costs measured in conflict. Implicit in that definition of choice is that there is no time left for learning.

That assumption may not be valid. Creating a sustainable fishery on the Columbia would evade tragic choice because conservation and harvest would be feasible together.[3] At our current level of ignorance, however, we do not know if such an evasion is possible. So, like physicians in an epidemic, managers must choose between heroic measures to save individual patients—which may be unavailing—and investing in a search for a cure for all—which is also uncertain. Under the Endangered Species Act, however, those stocks most in trouble are accorded priority; benefiting the already-sick comes ahead of searching for a cure. The Columbia has seen many stocks of salmon go extinct over the past century. That is no reason to value less the ones now in danger, but the earlier loss *is* a reason to value the capability to learn. To the degree that a recovery plan in the Columbia undermines the web of cooperation that supports learning there, the pursuit of a sustainable fishery is in jeopardy.

The problem is generic. So long as society values biologically unique species in general, fundamental conflicts between those values and the industrial order can only be postponed. Later choices may well entail "a nightmare of justice in which the assertion of any right involves a further wrong." That makes a disciplined search for sustainable use important, but the intensification of conflict also makes the search more difficult.

Policy-Oriented Learning

That conflict and learning have positive as well as negative synergies is the implication of another body of work on policymaking. Studies of American government and public policy illuminate the second issue raised by Dewey, the role of the expert. These studies contradict the assumption that experts are neutral, nonpartisan specialists. Instead, the contemporary picture of the policy process is closer to Adam Smith's hidden hand: policy and learning are the by-products of competition among policy actors, including experts, politicians, and bureaucrats, all of whom act as advocates.

Fundamental to this change of perspective is the abandonment of the concept of the public interest. If no one really is a steward of the public interest, then the playing field is level: no one is more legitimate than anyone else, even though different institutional positions still constitute different roles. Only a senator can vote on the floor of the Senate, but his or her statements—drafted by a staffer from materials provided by a lobbyist or bureaucrat—are no more likely to be an articulation of the public good than is a statement by any other partisan.

This is not simply a cynical view, although it inevitably implies a skeptical interpretation of the ceremonies of public life. Congress has repeatedly set deadlines for environmental compliance in pollution control and toxic chemicals, only to have them repeatedly missed. Arguments over new deadlines, after this experience, are not so much debates about compliance as they are encounters to make or defend reputations. No one can say what will happen during implementation, but one can know who won the fight over the language in a regulation or bill.

An interesting consequence is that learning becomes necessary; ideas matter in the process that one scholar calls "policy-oriented learning."[4] Policy-oriented learning results from attempts to win support for a policy position in a series of arenas: first within an advocacy coalition—one's own agency or interest group—and later against opposition in a legislative or administrative forum. Because of the social circumstances under which policy advocacy is carried out, the debate is often analytic in character, and positions are accepted or dismissed when advocates convince one another of the soundness of positions, not simply because one side has enough power to prevail.

The critical social circumstance is that policy debates are *not* political, in the way in which that term is usually understood. Most policy debates take place far from the public eye, within relatively small policy subsystems, networks of congressional committees, government agency offices, and interest groups that work on a single policy problem over long periods. The Columbia basin is a policy subsystem: a broad spectrum of interests participates—federal agencies operating dams and selling hydropower, state agencies and Indian tribes managing salmon, and citizen groups and businesses with economic or environmental interests. All these organizations have permanent interests in the Columbia basin, even though the access and influence of the actors may fluctuate over time, as has been the case for the Indian tribes. Although the members of the subsystem are often opponents on some matters and allies on others, they share two attributes. They understand the policies being made and implemented in the Columbia, and they spend considerable time and effort working to change those policies to suit their own needs. Moreover, when those in the subsystem reach consensus, policies change. Such a network can exercise what one observer of American politics called "majority rule by the minority that cares." But a policy subsystem is not often visible in the mass media, nor are its issues often contested in elections. Even though there may be rivalry and competition, the politics of a policy subsystem are rarely democratic, in the sense of being understood or decided by voters. The policy subsystem is the world of the policy wonk, and the "minority that cares" may be quite unrepresentative of the populace whose taxes support a given policy.

Yet precisely because it is both small enough for communication to be complex and for its members to pay attention, the policy subsystem is also where learning can and does occur. Debates are frequently scientific in form: research is sponsored to gather evidence or to analyze existing information; the data are challenged before a community of professionals; and disputes over data or analysis are decided on the basis of articulated arguments and calculations, whose worth is settled by informal consensus. Those who do well in analytic debates persuade their colleagues and influence policy.

Learning is achieved through debates that cross lines of advocacy, changing the positions of groups and sometimes altering the character of conflict among them. The key weapon of the Northwest Power Planning Council in stimulating energy conservation was a surcharge on the electric power rates charged by the Bonneville Power Administration. The council's right to impose that surcharge was challenged by homebuilders, who filed suit, arguing that the council, whose members were appointed by state governors, could not direct BPA, a federal agency. Raising that question of federal supremacy under the Constitution shifted the grounds of the debate far beyond energy efficiency. When the U.S. court of appeals upheld the council's powers, much more was settled than energy efficiency: the ability of the states, through the council, to influence BPA was clarified and strengthened. As this example suggests, debate in a policy subsystem can be conducted before official bodies such as the federal courts, and the debate may not appear to be scientific. But it is a debate in which, as Dewey suggested, a proposal for action can be treated as a working hypothesis, tested by extended argument and tried against experience.

Indeed, policy-oriented learning has occurred in a most unpromising setting, the process of writing environmental impact statements to comply with the National Environmental Policy Act of 1969. The policy was set forth in legislation only superficially, and its meaning was worked out in courts, as a series of lawsuits produced rulings that compelled federal agencies to change behavior. Even under these circumstances—which hardly seem to favor a rational procedure—environmental debate in the federal agencies has been made routine. The agencies must disclose information and analysis, and, according to one thoughtful observer, although "the courts do not provide an acknowledged forum in which technical issues can be definitively resolved . . . they may provide enough of a forum, *de facto*, to keep the agencies on their toes." The result has been an improvement in the environmental profile of federal activities.

Yet the reach of policy-oriented learning is limited, as political scientist John Kingdon suggests in an influential book on the way policy problems emerge in government. His first conclusion is surprising mainly to social scientists: elected officials matter. They set the agenda of government by determining what they themselves will and will not pay attention to. Setting the agenda is not the same thing as making the choices that appear on the agenda, nor is it the same as controlling the alternatives considered in the decisions. But defining what large organizations will and will not attend to is both important and plausibly under the control of those at the top of the hierarchy. Their discretion is not total, of course, but the complex world faced by any large institution necessitates sharp simplifications. Attention is a scarce resource; it is allocated mostly from the top.

The implication for conflict and learning is that occasions for either are limited. There is ample opportunity for disconnections, situations in which the effects of one choice are not linked to the context of other choices. But disconnection also sets priorities, though inadvertently. The fact that elected officials' priorities count suggests that they can be held to account for them. Yet as political scientists have shown for a generation, there is no sign that American elections produce either coherent policy mandates or accountability for such mandates.

Kingdon's most startling and persuasive claim is that the policy process is not a process at all in the usual sense; rather than being a sequence in which information is assembled and debated and decisions reached, Kingdon argues, government is better understood as an organized anarchy. Unlike models based on a mechanical conception of process, the idea of organized anarchy is biochemical: instead of conveyer belts on an assembly line, there is a "policy primeval soup," a loosely coupled set of actors, ideas, and institutions. "Generating alternatives and proposals in this community resembles a process of biological natural selection. Much as molecules floated around in what biologists call the 'primeval soup' before life came into being, so ideas float around in these communities." There are forces that move things along: elections, budgets, and other dead-lines to which actors respond. But what happens in detail is more a matter of simultaneity than one of causality: *when* an idea was advanced as a possible solution or which meeting a person went to can affect what happens more than logic, interest, or power. The organized anarchy gives coherent expression to the widely shared impression that timing and luck matter a lot in large institutions.

Both conflict and change, including learning, contain important random elements. Large changes and surprises are possible, even likely, because *every* outcome is *un*likely: "a subject rather suddenly 'hits,' 'catches on,' or 'takes off.'" Large organizations still do not change quickly, because the alternatives may be few and discretion-ary movement is limited by considerations of budget and personnel. But it does mean that the focus of attention can shift rapidly and surprisingly. Thus, opportunities for learning are present. Percep-tions can change quickly, and ideas matter. Turning ideas into last-ing policies is expectably slower, because of the inertia of activities already under way. Thus the changes in the Columbia basin look both drastic on a time scale of decades, and insufficient when mea-sured by the perilous hold on life of endangered salmon runs.

Imperfect but Real Learning

The agenda mapped out by Dewey's hopes for an informed citizenry can now be filled in.

- In the making of policy, citizens at large play no significant role, except in the form of uncertainty over which elected officials will be returned to office.

- Experts play an important role *as advocates*. Conflict between advocates drives learning.

- Policy subsystems keep attention focused on problem areas for long periods. Citizens, the media, and elected officials do not.

- Some citizens do support experts through interest groups, and interest groups representing citizens play important roles in some policy subsystems, including the environment.

- But the idea of experts as teachers of the citizenry receives no support at all.

Learning occurs, ideas matter; citizens and democracy in a direct sense do not. That the society is democratic in its political culture is important, however, because democratic idealism maintains the openness of the policy subsystems both to newcomers and to occasional scrutiny. But connection of the subsystems to one another in a coherent national strategy—a task that should in principle be done by all political candidates and their parties—is achieved haphazardly by the budget process, not on the basis of ideology or rationale. Whether there is enough conflict to force learning may vary by policy subsystem, although this phenomenon has not been studied systematically.

But the greatest difference between Dewey's world and ours is the shift in *gestalt*: from a mechanical, rational, but venal world of self-interested political actors bound by outmoded ideas, to a fluid, anarchic world of professionals, unmoored from the voters, seeking ideas that will solve problems, many of which lack clear outlines or solution. There are conflict and learning in both worlds, but they are different worlds.

Reliability

If political conflict can produce learning only sporadically, is there a more reliable solution? The idea that intelligent design of institutions can mitigate the effects of human error and selfishness is at least as old as the Constitution's arrangement of checks and balances. I discuss two variations here.

When the individual parts of a system are prone to failure, the prudent designer provides backups, redundant components that will increase reliability or decrease the costs of failure: the second parachute packed by the skydiver, the tie I keep in my office closet for unexpected important visitors, the phantom reservation for the wary traveler—and the overbooking done by the airline—each of these is a redundancy meant to protect against a worrisome occurrence.[5] The parallel in American government is the sharing of governing authorities among different branches of government with different loyalties to voters and the state. By separating the powers to govern, the writers of the Constitution sought deliberately to create a government that would function better than the people who made it up.

Ideas about reliability help to define what is wanted from conflict and learning. The pursuit of sustainable development does not need a policy system that will invariably produce a particular end result;

the point of learning is that we do not yet know what the end result should be. What is needed, though, is a way of fostering social learning that will not bog down. Redundancy may be especially valuable when one needs to catch errors that are not correctable, or when change requires detailed alternatives. Both conditions arise frequently in ecosystem management.

Deliberately fostering diversity—learning in many different places, each supported in a different way, each pursuing its own mix of activities—enhances the chance that important lessons can be learned somewhere. The multiplicity of government agencies, Indian tribes, and interest groups in the Columbia basin can thus be seen as an advantage, since the failure of one or more to contribute to the adaptive approach need not doom the whole enterprise. Conversely, dependence on a single large source of funding from power sales is a point of vulnerability.

A contrasting instance is the environmental impact statement (EIS) procedure under the National Environmental Policy Act. Emerging through litigation rather than through explicit design, the EIS process was shaped by judicial authority and propelled by many different localized disputes. The process developed into a powerful tool for short-term learning.[6]

That EIS policy grew out of judicial interpretation affected the way the policy was enforced. There was no administrative penalty from within the executive branch to agencies that did not comply; instead the penalty took the form of unpredictable delays. Delays caused political embarrassment. Although administrative enforcement of a policy of environmental assessment might have been more precise, in fact the shift of organizational culture necessary to take environmental impacts seriously was so large that judicial fiat was probably no less effective. Not knowing which judge might sign the next restraining order, agencies responded more strongly than they might have done to administrative enforcement, which would have been negotiated with organizational superiors and would have been constrained more tightly by budget ceilings and personnel limitations.

The legal approach also gave formidable tactical advantages to opponents. An impact statement satisfactory to the courts had to meet procedural requirements: what mattered was not whether there was an environmental impact but whether the agency disclosed all relevant information about prospective impacts, together with a plausible demonstration that it had developed and analyzed all the information it needed to reach a reasonable conclusion. Environmental data were unfamiliar to agency bureaucrats, and it was unclear how much had to be gathered or how they had to be analyzed. At each step, opponents—who often did not care whether the analysis was adequate but only wanted to stop a project—offered legal challenges. Often enough, they were successful.

This situation put the burden of data collection on proponents, forcing them to hire knowledgeable analysts. Their effectiveness in changing the way environmental factors were taken into account was amplified by two forces beyond the courts. The first was the growth of the environmental movement: a set of interest groups with members who cheerfully underwrote the filing of lawsuits. The second was the existence of several federal agencies that were either authorized or required to comment on draft statements. Together, professionals in the other agencies, activists threatening to sue, and judges who were unpredictable if a suit were filed constituted an accidental but powerful force for policy-relevant analysis—accidental in design but redundant in the way in which they reinforced one another. Change came with startling speed, building momentum even under administrations with little sympathy for environmentalism.

Conflict through the courts forced substantial change in the agencies' decision making. But there was little learning about the environment itself. The reasons lay in the same incentives that made the federal agencies unwilling agents for the environmentalists. Environmentalists seldom wanted a better project; they wanted no project at all. Environmental assessment was thus a pretext, often a successful one, for opposition. But win or lose, both proponents in the agencies and their constituency groups and opponents in the national environmental movement moved on. Cancellation of a project ended the matter. Approval also ended the environmentalists' concern about it: they had lost, there were other battles and insufficient resources, and the projects were usually a long way from Washington, D.C., where the battles were fought. It was a dynamic that forestalled learning about the environment, although it forced rapid learning in the agencies about what had to be done to comply. What remains is environmental analysis that is often usable, but with few users and little cumulative ecological learning.[7]

Negotiation and Planning

What the theories of public policy say, in sum, is that institutional networks channel conflict in ways that sometimes promote learning, that reliability can sometimes be enhanced, and that the whole system is sufficiently fluid and anarchic that better solutions may not be feasible. An implication is that informal processes, operating in situations of conflict and institutional disarray, can make useful contributions.

Since the mid-1970s there has been remarkable innovation in informal dispute resolution. A surprising blend of technocracy and negotiation has gained visibility and favor among managers of natural resources. In cases previously characterized by lengthy litigation and embittered conflict, informal negotiations have produced plans of action acceptable to traditional adversaries: tribes and state governments, environmentalists and developers, resource managers and harvesters. Though wary of advocacy in the guise of science, the parties have found it possible to use technical analyses and have invented measures to assure the political and scientific credibility of analysts and their findings. The negotiated agreements have sought to achieve and maintain the measure of consensus necessary for experiential learning to occur.

The task of these negotiations is to work out "rules of the road," ways to *continue* disputes within a process that all parties regard as workable. The negotiations are *not* aimed at ending conflict or politics, but at restructuring it; as one analyst put it, "Environmental mediation is best seen not as an alternative to environmental politics as usual, but as a strange new form of it." Developing rules of the road entails two tasks, organizing and planning.

Organizing

The political task that precedes negotiation is to organize contending parties so that each is able to deal with the others. Environmental disputes are characteristically many-sided; in most places there are multiple users of resources and multiple claimants to rights. Each user or claimant stands in a distinctive material and moral position with respect to the issues at stake; the participants will tend not to cohere into only two positions. While Indian tribes and state fisheries managers both want more water flowing in the Columbia, they remain rivals for the harvest.

The complexity of multilateral negotiation is often compounded by the difficulty of finding someone able to speak authoritatively for each of the parties. Environmentalists' concerns usually arise defensively, and there is no long-standing organizational structure whose leadership can commit its members; government agencies endanger the legal myth of sovereignty by admitting that they are parties at interest—even when it is clear to all that they are far from being neutral magistrates. In this context, the creation of a negotiating framework entails a nontrivial effort to organize the parties so that they can enter into some relationship with one another beyond that of opposition.

For this reason a "third-party" intervener can play an important role, helping the parties to organize and facilitating negotiation among them thereafter. Environmental dispute resolution developed along several different lines, and the interveners' approaches emphasized different visions of conflict and led to different notions of how learning from experience should proceed.

Decisions without agreement. Conflicts end in three ways: one party may lose—outright or by unilaterally withdrawing or avoiding a dispute; the contending parties can negotiate; or a third party can intervene to improve communication, facilitate negotiation, or render a decision (either advisory or binding).

Third-party intervention is an attempt to alter decision-making processes so that disputing parties can make choices jointly or consider how and whether they should do so. A typology developed by sociologists James Thompson and Arthur Tuden usefully summarizes variations among social decision processes. They observed that choices differ in two important ways. First, there may be agreement or disagreement about causation—about what will happen if one of the decision alternatives is chosen. Second, there can be agreement or disagreement over which outcomes are preferred. Put simply, decision-making situations may find those involved differing with one another over means (causation) or ends (preferences). Thompson and Tuden suggested that each combination of agreement or disagreement requires a distinctive decision strategy and organizational structure for making choices (see Fig. 4–1).

When there is agreement on both causation and preferences, decisions are neither controversial nor conceptually difficult. Decision making under these conditions can be routinized: choices can be made on the basis of orderly processes for gathering and evaluating information. The appropriate organizational form for routinized decision making is *bureaucracy*, the characteristic form of executive agencies. Conversely, public policies implemented by bureaucracies implicitly *assume* the absence of conflict; the presumption is part of the reason bureaucracies are readily stymied by disputes. Public servants are taught to serve the public, and they are puzzled when voices, each plausibly claiming to represent the public, offer contradictory reactions to how they are being served.

When agreement over the way things work is weak, *collegial judgment* is needed. Collaborative weighing of evidence is found both among scientists searching for the most probable version of the truth and among jurists seeking just interpretations of law. Collegial judgment relies upon *consensus among experts* to win acceptance in the wider society.

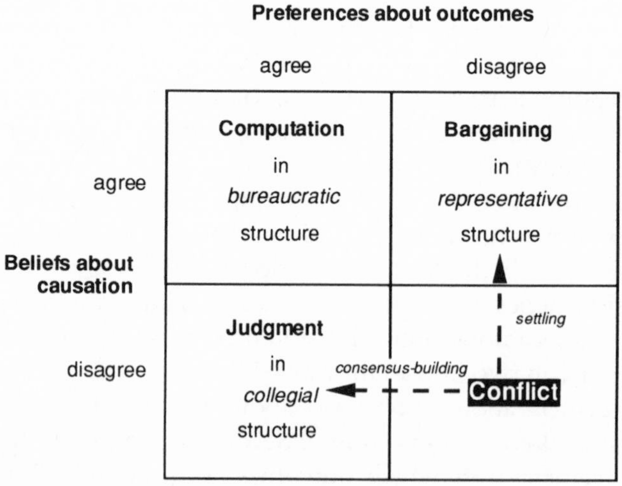

Preferences about outcomes

FIGURE 4–1. Decision making under varying conditions of agreement, and two strategies of intervention (after Thompson and Tuden 1959).

When people disagree about preferences rather than about causation, the decision-making style must harmonize competing ends through compromise and *bargaining*. In the legislative process, for example, elected representatives must work together to represent the divergent concerns of their constituents. When there is disagreement about both ends and means, there may be no clear way to proceed; Thompson and Tuden suggested that in this situation, inspiration by a charismatic leader may be the only way to reach decisions.

For present purposes I want rather to characterize conflicts as situations in which disputes over preferences and causal beliefs become intertwined; hence the understanding of how the dispute is linked to means and ends also becomes problematic. As a conflict develops, associations are formed, among persons and between persons and ideas. Some means (and experts associated with their use) become identified with some ends (and their supporting coalitions). In addition, in the pragmatic culture of the United States, most public decision-making forums focus on specific projects and policies. The choices are thus cast as means toward ends, ends that are either taken for granted or else stated in abstract terms. As a result, conflict over means becomes a way of disputing goals. The confounding of ends and means leads to confusion and dispute about process, about how to decide. This happens in part because—as disputants understand—different decision processes lead to different outcomes. But another reason for the confusion is that it is not evident *how* choices of decision-making processes can affect the levels of agreement and disagreement. Perspective on these uncertainties of process is one of the important services an outsider can provide to disputing parties.

Such advice is not unbiased, however. How one analyzes a conflict is affected by one's views on—and interest in—intervening. Practitioners whose understanding of environmental conflict is rooted in science and planning have a tendency to think that solutions can be found if only the parties will defer their animosity long enough to see how they can all gain. Practitioners whose training is in labor relations tend to think that the opposition of interests is permanent, and that the objective should be to find out whether *any* joint actions are possible, rather than trying to pursue harmonious resolution. Interveners from both backgrounds have entered environmental disputes—and clashed with one another about the correct view of conflict.

Yet in practice, the distinction between means and ends *is* unclear. Most ends are not final goals, but proximate objectives leading toward a more distant outcome: near-term ends are also means. Figure 4–1 is important, then, not because it relies upon a distinction that is clear in fact, but because it can be used to identify two conceptually distinct strategies for intervening in a conflict: either attempting to move toward agreement on causation, leaving preferences to be reconciled later; or attempting to move first on preferences.

The first approach to intervention seeks to obtain agreement on preferred outcomes, so that questions of means can be addressed by parties jointly. This intervention strategy is often called *consensus-building*. It is normally attempted as a prelude to joint planning, in which parties try to turn their initial consensus on goals into a plan that all can endorse. The preferences parties can initially agree to tend to be distant abstractions—such as "sustainable development"—since they are already wary about each other's motives on concrete results in the near term. This strategy is, in general, feasible only when a major change is introduced from outside, so that new issues and new alignments of parties at interest can supplant existing lines of division. There may be a high potential for conflict, and great concern at the pace at which conflict will unfold, together with uncertainty about its social dynamics.[8] As potential conflict becomes actual, goodwill between the parties is eroded, and the idea of agreeing on distant goals becomes less attractive: there is less to be gained from communicating with one's adversaries, and those who think themselves vulnerable (possibly all parties) are suspicious that agreeing on goals is a snare or delusion.

A second approach seeks to alter the character of the dispute by obtaining agreement on causation, deferring questions of ends. This strategy launches a process of bargaining and negotiation, usually by representatives of larger groups or interests. This intervention method may be called *settling*, since the aim of the negotiation is not to achieve final resolution of conflict, but rather to hammer out joint actions within a relationship in which all parties are aware of and retain opposed interests. The reason parties whose interests are opposed to one another—management and labor, polluter and regulator—nonetheless try to work together is that they have to. Their agreements tend accordingly to be forced by deadlines, to be short-term and limited in scope—in a word, to be contracts rather than joint plans.

Both settling and consensus-building create *negotiating processes*, in which disputes may persist among parties. Since the agreement to negotiate is normally voluntary, there are few sanctions other than the negotiating relationship and what it produces to discipline the parties. An agreement, from this point of view, is a point of departure, not a destination.

Learning. Persisting differences among disputing parties affect what is likely to be agreed to and the role of learning within the agreements. In this context, the notion of learning brings us back to Dewey's claim that the main need of the public is for "improvement of the methods and conditions of debate, discussion and persuasion."

Consensus-building focuses on shared objectives and urges the sharing of information as a sign of trust. The danger to parties is that some will not honor this trust and will use the information to twist agreement on abstract goals into consent for unpleasant realities. A structural safeguard that reduces this hazard but does not eliminate it is public disclosure of information by all parties. Those with the most to gain by developing projects also know the most about the site, its utility for their purposes, and the risks of the project. Their commitment to a shared approach to planning can be tested early and late by their willingness to disclose information. Conversely, those who seek to prevent development are pressed by an open-disclosure framework to provide reasons and evidence for their opposition: if they share objectives, at some level, with those who want to modify the environment, they are obliged to discuss uncertainties openly and to abide by the conclusions implicit in information gathered jointly. Not all perceived risks stand up to scrutiny; having reached a consensus on goals, the opponents are yoked to the results of investigation.

Settlements aim at learning for the purpose of enforcement, to assure that other parties are following their obligations under the agreement. Their joint risk is that in protecting their individual stakes they will overlook or fail to achieve better outcomes for all.[9] In these situations too, open disclosure provides an opportunity to discover a superior outcome. But that may not be enough: a superior outcome is often a public good, with some value to each party, but with not enough value to warrant the risk of being put at a disadvantage by one's adversaries.

There are two conceptual solutions. One is for parties to build cooperation gradually and deliberately. Building cooperation requires consistency of behavior, something that may be difficult for parties that are themselves coalitions or organizations. The alternative is to develop an institutional structure in which the execution of the agreement is in the hands of an independent party, which becomes a natural advocate of superior solutions. This is what the Northwest Power Planning Council became in the Columbia basin. Developing such a structure, however, is seldom a low-cost solution, particularly if the agreement is to be of short duration.

Asking what can be learned from a negotiated agreement suggests two points. The first is that Dewey's analysis might be appropriate after all: experts brought into a negotiated agreement can be neutral interpreters because their mandate is to work for a coalition of interests that spans the spectrum. Moreover, if the negotiation process works well enough to keep the disputants involved, the parties will also be teaching each other about their different goals. This is a precondition for developing shared goals that mean something concrete.

What makes Dewey's model plausible here, while it is not in the national policy arena, is a difference in scale. Dispute resolution is a face-to-face process in two senses. Negotiators always meet one another in small groups. But in attempts at informal dispute resolution, each negotiator also has a face-to-face link with his or her constituency, the group the negotiator represents. This is usually not the case in national policy debates, in which the constituencies are far removed from the "policy soup" and there is usually no practical way for negotiators to educate constituents via dialogue. Negotiators' latitude is constrained by what their constituencies will ratify. When it is possible for the negotiator to "bring along" his or her constituency by educating its members—a means of defining the limits of the negotiator's discretion—the encounter *between* negotiators can reach ground that would otherwise be inaccessible.

A second observation is that consensus-building is likely to be insufficient by itself. Learning is a gradual ascent toward confidence punctuated by the slippery panic of disappointments. If divergent goals are papered over by agreement on abstractions, each disappointment is a moment when good faith can evaporate. In the more suspicion-laden world of settlements, in which expectations are lower, it is prudent to design ways at the outset for sorting out interpretations of disappointing results. A consensus-building process that makes such an arrangement has gone a substantial distance toward embracing the settlement process and will be stronger for it. The scientific method of adaptive management is such an arrangement, testing expectations against outcomes and providing an evaluative framework in which surprising outcomes are expected rather than demoralizing.

Planning Centralizes, Implementation Decentralizes

Once a framework for continuing negotiation is in place, it is possible for parties to work out the details, a process I shall call "planning" even when the character of the agreement is a settlement. Planning is the assembly of information and analytic skills able to describe the world shared by the parties and able to identify the uncertain consequences of action within it.

Planning of this kind proceeds both when the parties agree on preferences and when they do not. Without a framework for disputing, there is no social mechanism for exploring common objectives, and no means to reach joint commitments. Inquiry foreshortens to intelligence gathering. Decisions are the residue of conflict and evolving circumstance. Without organized analysis, knowledge degenerates into advocacy—their experts versus ours. Inquiry is then yoked to competing objectives. Learning is the accidental residue of conflict, and hard-won lessons cannot become the shared property of disputants. By contrast, planning in a framework of active negotiation is a process in which conflict can proceed productively, permitting and encouraging the articulation and defense of contending values.

That process was used by the Northwest Power Planning Council. Backed by a legal mandate to provide public information and involvement, the council sought to lower barriers to participation and to judge its success by its credibility with the public. From its first years on, the council approached organizational and opinion leaders both inside and outside government and consciously built a consensus to back its energy plan and fish and wildlife program.[10]

The link between negotiation and adaptive management can be

seen as a reflection of a counterintuitive but crucial proposition: learning requires conflict. Because planning centralizes knowledge, it is natural to assume that planning works best in centralized, disciplined settings in which goals are agreed to. But an adaptive search for sustainability learns from implementation, and implementation decentralizes power. The tension between planning and implementation needs to be managed in the pursuit of sustainable development because implementation supplies the signals of what does not work, what needs to be modified in a centrally developed plan. This general tendency is made explicit in adaptive management.

The natural systems being managed cross the spatial and functional boundaries of existing institutions. Without a comprehensive perspective, the fragmentation of jurisdictions promotes unsustainable abuse of the environment, because individual institutions seek to achieve purposes that often turn out to be incompatible with sustainable use of the whole. Yet implementation requires a decentralized, fragmented perspective, because decisions are carried out by parties whose responsibilities are narrower than the analytic tools used by analysts.

Political competition. Managing the tension between planning and implementation entails, paradoxically, the fostering of political competition. A competitive environment does not emerge automatically, but planners can use negotiation and analysis to link the agendas of separately managed organizations. Planners should not see themselves as managers of the organizations or as superordinate powers. Instead, they should aim at the internalization of both political and economic costs, so that the forces of error correction can be marshaled when surprise and error emerge.[11]

In the Columbia basin, for example, fishery enhancement is funded by hydropower revenues. This arrangement is desirable on economic grounds, because those benefiting from electricity help to pay for the damage its production has caused. In fact the impact on electric power rates is small. Nonetheless, utilities and large users such as aluminum refiners feel an organizational and political incentive to act as watchdogs, seeking to ensure that fisheries managers achieve results. Thus a political mechanism reinforces an economic means of recognizing error. Such an institutional design builds a social infrastructure for adaptive management.

Political competition is also important because it increases the likelihood of durability. Once established, a competitive political order is generally difficult to remove; political competition resembles the introduction of biological species, which is difficult to reverse once established. Encouraging political competition improves the likelihood that error correction will continue.

Inevitable conflict. In social learning, conflict is inevitable.[12] The challenge is to keep conflict within bounds. The agenda of any dispute, the definition of legitimate disputants ("standing" in a legal setting), and the procedures—including production of information and presentation of arguments—by which the dispute is conducted are all governed by institutional arrangements and the distribution of resources. Changing these factors, piecemeal or strategically, to facilitate coherent outcomes is the institutional challenge of social learning.

Tension between centralized knowledge and control and decentralized experience is perennial. Central planners and governments have a wide-angle but abstract view. Because they must deal with diverse and complex aggregations, there is a constant temptation to oversimplify. To managers at the center, it is imperative to channel the flood of information to manageable proportions. The task is to focus the information received in a way that produces lasting solutions to real problems at a tolerable level of discord. Tradeoffs appear at the system level that are incomprehensible or unjust from the local perspective. Whenever local preferences matter, tradeoffs are constrained, and there may be losses to the whole—losses that result in barely perceptible overall erosion or in a shifting of damage from one place to another.

Conversely, the way the center sees things is bound to seem incomplete, unfair, or ignorant to one or more observers at the local level, whose experience is firsthand, and whose responsibility and information do not include factors important to overall stability and other system-level properties. Because local knowledge is more detailed, often more persuasive, and sometimes contrary to the inclinations of those at the center, local actors find themselves in the position of forcing the center to attend to matters ignored, to integrate values separated by abstract understandings of what is needed, to change the center's grasp of what an action means or the rules by which a decision should be taken. This fosters conflict. It is not the kind of conflict that is readily handled by markets or anticipated in contracts.

The tension between centralized planning and decentralized implementation also reflects the difference between learning to perform a task better and learning to revise objectives in light of experience. In the mitigation efforts in the Columbia basin, two kinds of experience accumulate. The first is how to carry out a particular kind of project, such as building a screen to guide juvenile salmon safely through a dam, bypassing the power turbines. This sort of experience can lead to lower costs and higher efficiencies. A second kind of experience develops at the system level, as different kinds of projects are implemented and their effectiveness compared and evaluated. System learning tends to develop in the centralized elements of the program; task-based learning develops at the local level. Both are important to the search for sustainability: task-based learning is important to carrying out projects in cost-effective ways; system knowledge is essential to improving strategies in light of experience.

The complexity of human and natural systems is high enough to outrun anyone's ability to plan entirely from a central vantage. A framework for conflict and negotiation, designed with foresight and operated with understanding, is necessary to link centralized perception with decentralized action.

Social learning is a creature of politics and institutions. It is uncontroversial to seek to learn from experience. But in order to discipline and to speed that learning with the experimental method, it is necessary to maintain a level of institutional cooperation and stability capable of planning, implementing, and monitoring activities over the length and breadth of large ecosystems, and to persevere for biologically relevant periods. Controversy is essential for error recognition to work, but severe conflict can undermine the error-correcting capability itself. In responding to controversy, it is essential for political leaders to understand both ends of that proposition.

Learning spurred by conflict can occur in public policy, and learning can be stimulated in informal agreements among disputing parties. But it is not assured; conflict is a supplement to adaptive management, not a substitute for it. In any event, nature is sure to provide surprises, some of which are likely to spur timely learning.

Chapter 5

Sea Trials: Comparison Cases

Science is experience becoming rational. The effect of
science is thus to change men's idea of the nature and
inherent possibilities of experience.

—John Dewey, *Democracy and Education*

Social learning comes from the accumulation of knowledge within a
network of organizations and from conflict between organizations
and their environments. In the last three chapters I have described
these two complementary modes of learning. Both can produce
changes of behavior and understanding within an institution, although there are also many ways in which learning can be thwarted.
In this chapter and the next I focus these concepts in two different
ways. Here I consider other situations in which learning and ecosystem management have been tried—cases from the Australian rain
forest, fisheries management in Canada, timber harvest in the Pacific
Northwest, international institutions including the management of
pollution in the Mediterranean, and criminal justice and health care.
Experience in other places and other cases illuminates the idea of
social learning in the arc light of reality. Then, in Chapter 6, I merge
theory and cases into a discussion of how social learning can be
fostered and why policy cannot guarantee that learning will occur.

Ecosystem Management

Three instances of resource management at the ecosystem scale, in
Australia, the United States, and Canada, yield cautionary tales and
some useful positive notions about the link between organizational
structure and learning.

Queensland Tropical Forest

In the 1980s, under pressure from Australian environmentalists and scientists throughout the world to "save the rain forest," managers of the wet tropical forest of northern Queensland, Australia, attempted to fashion a sustainable management regime for its timber. This plan is discussed in *No Timber without Trees*, a worldwide study of sustainable forestry by a team from the International Institute for Environment and Development (IIED) in London. The team was led by Duncan Poore, a forestry scientist at Oxford University.

The Queensland experience confirms the importance of combining science and policy, in a mix similar to that in the Columbia basin. Because the biology and economics of the tropical forest in Australia are much closer to those of developing countries, Queensland is a bridge from the industrialized Columbia to the setting in which sustainable development is being urged. Yet the Australian forest experience also harbors a warning: Queensland's forest land was withdrawn from production in 1987. The land was declared a World Heritage site, and logging is no longer allowed there. That a promising attempt at sustainable management was derailed by the politics and science of postindustrialism is a striking omen.

The rain forests of northern Queensland grow on high plateaus, at the limits of the tropical rain forest climate. The rain forest covers an area of just over 1 million hectares, or about twice the size of a national forest in the continental United States. The soils support approximately 800 species of trees, 150 of which are harvested for wood. By contrast, the New England states of the American Northeast produce 17 species of commercially harvested trees. The dominant tree species in Queensland are well adapted to disturbance and regenerate after cutting, although controversy remains about whether species are lost by logging. The Australian rain forest does seem more tolerant of human intervention than most other tropical ecosystems, a virtue of considerable importance if experiments are to be conducted.

In the 1950s there was enough virgin forest—what is now called "old-growth forest"—that it seemed inexhaustible. There was never agreement on the size of the sustainable yield, or even if the idea of sustainability—of harvesting a volume of wood equivalent to that which grew—made sense in a complex tropical ecosystem. But the final proposal of the Queensland Forestry Department, of 60 million cubic meters for 1988, would have been more than twice the entire harvest from the U.S. national forests, which cover an area more than 100 times larger than the Queensland rain forest.

Sustainable management plans. Nearly all of the Queensland rain forest is in public hands, controlled by the national and state governments. According to the IIED analysts, this land tenure promoted substantial learning about the commercial uses of the rain forest. Planning by the Queensland Forestry Department was based upon long-term studies of 37 growth and yield plots located in different parts of the forest. The plots included both logged and uncut lands, and contained half the 800 species of native rain forest trees; monitoring of these plots had gone on for periods of 10 to 35 years, with the aim of measuring the productivity of forest lands in a variety of conditions of use. This base of data provided control cases against which forest practices and other experimental treatments could be measured. But when controversy over logging erupted, an internationally known biologist dismissed the state government's research as "very improbable."

From 1945 through the early 1970s, experiments were carried out at 27 research plots, to study how planting and land treatments would affect productivity. Because the goal of these studies was higher economic output, they were not planned with ecological parameters in mind. The research did turn up an important generalization: rain forest hardwoods, the source of high-value veneers, did not respond well to the treatments, and the product from natural, unmanaged forests was superior. It therefore made economic sense to think of sustaining natural forest.

Because of the large number of species of trees, timber harvest plans designated specific trees for cutting, as well as the direction in which each tree was to fall (to avoid damage to other valuable trees nearby) and environmental regulations to be followed in the timber operations. The actual logging was carried out by private firms, operating under licenses from the government. Follow-up studies of logged lands gauged the effectiveness of the environmental regulations. Although about one-eighth of the neighboring trees were damaged by logging, the forest lands grew in again within 5 to 6 years, with a mature forest structure expected to be in place by the time of the next cutting 40 years later.

Under these arrangements, logging was profitable, and was projected to remain so. The projected volumes of harvest, which were much lower than historical levels, would have been sustainable indefinitely, assuming market prices remained stable. But in 1982 the wet tropical forest of north Queensland was identified by the International Union for the Conservation of Nature and Natural Resources as one of the world's greatest natural areas. That scientific judgment implied preservation of the rain forest. Not everyone agreed that logging had to stop, and the Queensland Forestry Department and the state government fought to save the timber industry there. Nevertheless, the environmentalists had the votes, both in the international World Heritage Committee and, more important, in the Australian parliament, which banned logging at the end of 1987.

However promising this sequence of events may be for the preservation of biodiversity in the rain forest, the closing of Queensland to logging put an end to one of the only technically explicit attempts at sustainable harvest. In the battle leading up to World Heritage designation, critics of the Queensland foresters, including the chair of their own scientific advisory committee, argued that the loggers had not done their science in good faith. Given the complexity of rain forest ecology, it is hard to appraise the merits of those criticisms. Without an experimental program of management at the ecosystem scale, however, we can be certain that some important questions will remain unanswered.

Principles. In its assessment the IIED team argued that the Queensland program, though stillborn, nonetheless identified critical conditions for sustainable management. Table 5–1 compares these conditions with those prevailing in the Columbia basin.

Like Queensland, the Columbia basin operates within a developed national economy. Neither is therefore an obvious socioeconomic model for use by developing countries. In both Australia and the United States pressure came as much from those arguing for preservation as from those advocating continued economic exploitation. In these respects Queensland differs from developing countries with tropical forests, which face pressures from expanding landless populations and foreign debt, sometimes exacerbated by corruption in government. These conditions prompted the leader of the IIED

TABLE 5–1. Comparison of Queensland management model and
circumstances in the Columbia basin (Source [Queensland program]: Poore et al.
1989, chap. 2).

Queensland Program Principles	Columbia Basin Situation
Secure tenure of the forest estate, so that long-range planning can be undertaken with the expectation of implementation.	Hydropower production and sales are assured. Until recently, the fishery was an "open-access" resource. Control of fresh water remains fragmented, so the salmon habitat is insecure.
Good *control of operations*.	Centralized coordination is excellent for power production, unreliable though improving for fisheries.
Profitability for loggers and timber merchants.	Stable revenues from the sale of hydropower undergird the economics of ecosystem management.
Research base to guide planning and environmental regulation.	Engineering base of the hydropower system is firmly established; there is a growing base of research supporting energy efficiency programs. Assembly of data and analysis are now under way for fisheries.
Cultural and social environment supporting *stewardship*. Research and management costs did not have to be recovered by logging. There is a long tradition of research linked to management planning.	Similar long tradition of management, social climate favorable to research, and bounded conflict among parties at interest.

team to remark: "It is evident that the future depends absolutely on the will and ability of countries with tropical forest to develop and apply effectively policies of sustainable and equitable land use, among them policies for the sustainable use of their forest resources. Any outside initiative should be judged by the extent to which it helps them to do so. The present cocktail of incentives and penalties, praise and strictures, which tends to flow from the better-endowed half of the world, is not a well-forged instrument for this purpose."

The Queensland experience suggests that the Columbia basin program is pointed in the right direction, but with limited transferability to other places and circumstances. The comparison between the cases brings out three points about the Columbia: the imbalance between the knowledge and economic bases of power production and of fisheries management; the narrow economic grounding of the salmon rebuilding program; and the decisive importance of environmental values (which remain fragile) among voters and the relevant levels of government. The Columbia basin's expenditures of $50 per salmon represent an investment in hope and justice, intangibles that a market is ill-equipped to assess. What the now-unemployed loggers of Queensland might have illuminated is the question of how hope can be anchored in economics.

Timber, Fish, and Wildlife

Forest practices on nonfederal lands in Washington State, in the American Pacific Northwest, are regulated by the Forest Practices Act of 1974. Intended to create an orderly regulatory process, the act became instead the focus of escalating conflict. By the mid-1980s, combatants had grown weary of political and legal battles that gained little for anyone, yet seemed necessary for everyone in order to protect their interests. Several parties began to seek other means to settle their differences. Their willingness to try face-to-face negotiation led to the Timber, Fish and Wildlife Agreement (TFW), concluded in February 1987 by representatives of 24 organizations, including state agencies, Indian tribes, the timber industry, environmental groups, and mediators from the Northwest Renewable Resources Center in Seattle. The "TFW cooperators," as they called themselves, were widely praised for bringing peace to one of the Northwest's battlefronts. The agreement revised the process for regulating and managing nonfederal forest lands in the state—lands in private ownership and state lands managed by the Washington Department of Natural Resources (DNR). Under the new regime, the timber industry has prospered, even though regulation of logging reflects a broader range of interests than in the past.[1]

The importance of TFW became increasingly apparent as disputes over federally owned lands in the Pacific Northwest made the spotted owl a *cause célèbre*, the symbol of an apparently intractable conflict between environment and economy. Yet in Washington State, on nonfederal lands dispersed among the national forests, cooperation reigned. TFW is a notable use of informal negotiation to search for a sustainable balance between exploitation and conservation.

It is unclear, however, whether cooperation will lead to better management. The TFW cooperators agreed that they needed more knowledge at the outset, and they even borrowed the term *adaptive management* from the Columbia basin program. But in two critical respects the regulatory regime and an accompanying research program fell short. First, neither regulation nor research was structured around watersheds, the most prominent ecosystems in the forest. TFW failed to see the biological wholes it sought to manage. Second, the cooperative research program had no channel for feedback to management and regulatory practice. There was no regular way for lessons inferred from science to affect implementation. For both reasons, the political commitment to learning is not likely to be realized until there is more conflict. Ironically, the promise of TFW was precisely that it would supersede conflict.

The TFW Agreement. Although TFW was not legally binding, the united force of the timber industry and environmentalists behind a single proposal led both the state legislature and state forest practices board to adopt the suggested changes. By 1988 the regulations governing forest practices bore the stamp of the TFW cooperators. These changes were backed by substantial resources from industry, government, Indian tribes, and environmental groups.

Additional components of the agreement outline an informal dispute resolution process. TFW defined a working procedure for making decisions on forest practices applications, most of which seek permission to cut timber. Roughly 10,000 applications were filed in 1988, covering the harvest of 4 million board feet of timber from private lands. Most of the permits were issued or denied on the basis of standardized criteria.

TFW also provided for on-site review of permit applications for lands of special sensitivity. These reviews, by so-called interdisciplinary teams, bring representatives of TFW cooperators to the site, where they negotiate with the landowner and with one another to determine the conditions to be applied to the permit. At this point tree-by-tree decisions can be made to conserve habitat or Indian cultural values while timber harvest proceeds. On-site reviews decentralize regulatory decisions to the forest floor.

Because there were long-standing disputes over the biological significance of forest practices, TFW called for a program of research. The studies are sponsored by a Cooperative Monitoring, Evaluation, and Research (CMER) Committee, jointly chaired by a scientist from the Northwest Indian Fisheries Commission and a scientific staff member of the Weyerhaeuser Company, a forest-products firm that is the state's largest private landowner.

The TFW agreement committed the parties to attempt to achieve the goals of all its signatories' organizations. The goals listed in the agreement covered wildlife, fish, water, archaeological and cultural values, and timber. These goals were a model of sustainable development: no timber without trees, but no trees without timber. In agreeing to these goals, each group explicitly accepted all other parties' goals as legitimate: "These needs are not mutually exclusive. They are compatible," the agreement declared. This optimistic declaration of unanimity is characteristic of the consensus-building approach described in Chapter 4. As noted there, consensus agreements without a procedure for interpreting surprises can be fragile.

Science and policy. If the diverse goals in the TFW consensus were all to be achieved, the problem-solving orientation of the negotiation process had to be sustained. As an analyst of environmental negotiation observed, "the parties involved in the dispute often have the facts and expertise necessary to make the most sound decisions; and ... this knowledge can better be tapped in open and cooperative negotiations than in adversarial approaches where information is usually hoarded." This characteristic of a negotiated consensus is especially important in environmental disputes, where the cost of information is a significant barrier. In that way consensus should produce better decisions.

Science got off to a good start in TFW. It became clear during the negotiations that the forest ecosystems of arid eastern Washington differ from those on the wetter slopes west of the Cascade Mountains. Yet regulations had been based on western Washington forests. All sides agreed to additional field studies, which resulted in modifications to the regulations.

After this bright beginning, progress slowed. CMER was established in 1987, and TFW's policy-level participants were aware of the importance of getting better scientific and technical information to policymakers, regulators, and field personnel. Yet three years later there was still no way for studies to inform decisions on forest practices.

A more basic flaw is the spatial template of the TFW agreement. Because negotiations focused on the actions of landowners, the regulatory reforms adopted property boundaries as their frame of reference. But the natural frame of biological concern is the watershed—the boundaries around which seed dispersal, climate variation, and habitat are organized. Although watershed planning was included as an alternative to detailed regulation, the planning approach is attractive primarily to large landowners whose property encompasses whole watersheds. Had the negotiators thought in terms of ecosystem management instead of in terms of harvest regulation, it would have been possible to organize large-scale experiments such as intensive logging within one watershed, with lands in a control drainage left unharvested while comparative measurements were taken. Under the current arrangement, with 10,000 permit applications coming in each year, there is no overview of the ecosystems being changed by harvest, and thus no coherent ecological basis for management or learning. TFW missed the opportunity to engage with the challenges of the *large* ecosystem with its multiple jurisdictions.

Nevertheless, TFW is an advance in environmental policy. Its shortcomings, which are emphasized here, mark how long the trail to sustainable development really is. The fact that the idea of sustainable forestry could receive positive publicity, generate larger budgets and philanthropic contributions, and put an end to litigation is one bright measure of how quickly a society of mass communications can move. But with its current superficial commitment to science and learning TFW has a long way to go before it can provide the notion of sustainability with the foundation of knowledge it requires—knowledge based on time that has biological significance in a forest.

Salmon Enhancement in Canada

The case of the Canadian Salmon Enhancement Program (SEP) affords insight into the internal dynamics of resource management agencies as they attempt to learn from and manage natural systems. SEP was established in 1977 within the Canadian Department of Fisheries and Oceans; its goal was to double the runs of salmon in British Columbia. SEP was supposed to reach this objective within a decade, at a projected cost of 300 million Canadian dollars. Motivated by an apparent decline in salmon abundance in British Columbia, the program also reflected growing confidence in several methods of fish culture, including hatcheries, artificial fertilization of lakes, and construction of human-made rivers known as spawning channels.

The national government provided most of the funding for SEP, with a small contribution from the province of British Columbia. In retrospect, technical staff consider the crucial factor in the program as the ability to persuade federal politicians to keep funds flowing: "we have the technology if you provide the money." Within SEP, however, staff recognized that there were also large uncertainties about the effectiveness of specific technologies at specific sites; accordingly, learning was an explicit part of the agenda from the outset. Monitoring and evaluation of projects enjoyed high priority, and approximately 10 percent of the SEP budget has been spent on monitoring, evaluation, and related activities. As part of an investigation of how learning actually occurs in natural resource management, Ray Hilborn studied the evaluation of spawning channels. Hilborn is a fisheries scientist who worked with C. S. Holling in the 1970s in creating the idea of adaptive management. Hilborn's account of the spawning channels provides a number of insights into the success of SEP in learning from nature.

A spawning channel is an artificial river with regulated flow, gravel size, and spawner density. Channels are normally constructed as a set of serpentine loops, bulldozed on a graded area, with gravel added, and some form of flow control structure to provide water. In some cases the spawning channels were built to compensate for degraded spawning habitat or in areas subject to flooding. In other areas channels were built simply to increase apparently underutilized rearing capacity in adjacent lakes. The major spawning channel projects in British Columbia have been for sockeye salmon. Sockeye spawn primarily in rivers in the fall, hatch in the spring, and normally spend a year feeding and growing in lakes. They then migrate to sea, where they spend two or three years before they return to the rivers, spawn, and die. The creation of spawning channels was based on the assumption that the production of additional hatchlings would increase the number of adults in the fishery. Six sockeye spawning channels, built in British Columbia in the 1960s and 1970s before SEP was created, are now operated by SEP.

Hilborn examined evaluation and response in three specific elements of decision making in spawning channel management. One involves channel operators' decisions about how many salmon to admit to a channel. If too few fish are allowed to spawn, some of the gravel area is wasted. If there are too many, overcrowding and disease can impair production. Channel operators operate control gates to permit entry of salmon into the channel, so they effectively control the number of spawners each fall. The following spring, operators can count the number of fry emerging from the gravel and traveling downstream to the lakes.

A second management decision involves the cleaning of gravel. Over time channel gravel accumulates sediments that block the flow of oxygen-rich water and reduce the survival of salmon eggs. Periodically the gravel needs to be cleaned, and numerous technologies have been tried, ranging from simply bulldozing and raking the gravel while water flows through the channel, to constructing special cleaning machines that inject air and water into the lower depths of the gravel and bubble sediment upward. The success of gravel cleaning can be assessed primarily by the survival rate of eggs in subsequent years.

A third decision involves determining, on the basis of past performance, when and where to build new spawning channels. In the late 1980s SEP built two new spawning channels. Hilborn asked how evaluation of the success of the previous channels was incorporated in the site selection process for the new channels, and in the economic analysis of their expected performance.

The spawner density and gravel cleaning decisions are made by small working groups at each channel site, and the results of each choice are known within a year. Experience at other spawning channel sites is used, so that the channel decisions are comparable, strengthening the learning process. Experimentation and learning played large roles in these two areas of decision making even though after more than 20 years of channel operation at six sites questions remain about optimum spawner density, and gravel cleaning technology is still being improved.

In contrast, there appeared to be almost no learning about the overall success of channels. In every case but one, the channels did not appear to produce additional adults in the first 10 years after construction. Now, after more than 20 years, one channel is clearly successful, two are apparently failures, and three seem to be modestly successful, although the evaluation for these three remains ambiguous.

Learning has been slow for several reasons. First, there is a long lag time; the life history of the fish dictates that adult production cannot be measured until five years after the eggs are deposited in the channel. Second, because there is considerable natural variability in the survival of salmon, many years must pass before an apparent increase in adult production can be confirmed. Finally, because the trend in most sockeye populations has been upward in the last 20 years, it is hard to tell if the increase in the abundance of adults in the rivers with spawning channels is due to the channels, or simply part of a coastwide trend.

But other problems besides those of life history and natural variability have impeded learning about the success of sockeye spawning channels. SEP has displayed an institutional reluctance to evaluate adult production. The first papers on this subject did not appear until 15 years after construction. Three of the channels have never been formally evaluated for adult production by SEP staff. The estimates of adult production to be expected from new channels built in the 1980s were 4.7 times higher than the observed production rates from the existing channels. Finally, the new channels were built in such a way that evaluation of their performance will be difficult or impossible. The channels built in the 1980s reflected political needs and engineering desires rather than biological evaluation of experience.

The SEP experience suggests what happens in the absence of conflict. Although the process of learning is well-aligned with the responsibilities of the operational organization, the information gained remains within the technical staff and makes no contribution to political and social judgments, apart from crude measures of whether a technology works at all. Social learning without controversy seems to be slowest at the system level.

Institutional Comparisons

Obviously, social learning is not limited to environmental issues or to problems that arise within a single nation. The cases described below involve social policy and international relations, two spheres of human endeavor in which learning from experience at large scale is of instrumental significance.

Social Experimentation

The confidence of the Kennedy presidency and Lyndon Johnson's Great Society was reflected in a feeling among influential social scientists that their work could have practical application. Many subscribed to the idea that public policy could be timed and scaled to steer the nation's economy. Systems analysis was then in high fashion, inspiring self-confidence among policy analysts and reinforcing the Great Society's hope that it was possible to go beyond coping with social ills and to solve them permanently. Early statements radiated confidence that experiments could be performed, useful and penetrating lessons learned, and policy improved. The decades since have brought disappointment and disillusionment. By the Reagan era a seasoned observer of social experiments concluded that few programs in human services, medicine, business, criminal justice, and education "have generated detectable and unambiguous effects."

The warning for sustainable development is a sobering one. The ravages of industrialism may be difficult to remedy, both because the damage is deep and because it is hard to learn what remedies succeed. The primary problem is to persuade taxpayers of the value of learning. Learning is largely a public good that must be paid for by government. If government becomes discredited by unfulfilled promises, willingness to spend for learning declines.

At least two other general lessons have emerged from social experimentation. The first is that learning at the operational and policy levels can occur under appropriate conditions. In a well-executed experiment in 1972–73, a study funded by the Police Foundation showed that preventive police patrols, which had been increasingly popular until then, did nothing to deter crime. Another early experiment, carried out from 1961 to 1964, showed that bail was not necessary for most people arrested for misdemeanors and some lesser felonies. On the basis of social background information such as employment history, stability of family, place of residence, and records of previous arrest, experimenters identified a population of those arrested who were likely to be acquitted later and who were highly likely to appear at trial. These were recommended for release on their own recognizance. The practice of releasing large numbers of those arrested without bail was subsequently ratified by federal legislation enacted in 1966, easing the burden on jails as well as on those arrested.

Both studies involved randomization, which enhanced their internal validity. Outcomes were measured in several ways, which agreed with one another and strengthened both internal and external validity. In both cases careful negotiations were necessary, with police officials and judges, to work out experimental methods that would be compatible with institutional routines and duties. And in both cases subsequent replication has affirmed the original findings, or at least not turned up surprises. These studies made a lasting difference in a field in which change has been forced more often by social crisis than by rational study.

A second lesson comes from the unfortunate reality that successful experimental learning is rare. The most persistent problem has been unwillingness to use random assignments to create control and treated populations. Randomization is sometimes not feasible: studies of accidents cannot assign potential victims to treatments, because their identity is not known in advance. Randomization almost always raises concerns about fairness. In his pioneering discussion of learning from experience, Donald Campbell coined the term *quasi-experiments* to designate studies in which control by randomization could not be used. In the way spelled out in Chapter 3, even in these cases threats to validity could be examined and their effects minimized by good design. As in the problem of statistical power, the advice sounded simple but proved hard to take. In 1986 a respected academic analyst warned afresh that "without extra assumptions and data, quasi-experiments are suggestive at best, and more often useless or downright misleading."

The implication for adaptive management is worth stating again: Controls and replication are indispensable, but they are expensive. There are two alternatives: well-defended quasi-experiments, in which challenges to validity are explicitly and comprehensively addressed; or misleading, unconvincing findings.

In sum, the small number of social experiments that have produced improvements in policy and operations furnishes warnings about the reliability of studies conducted outside a laboratory.

International Organizations

Sustainable development is inherently international because trade links national economies. In fact a logical approach to sustainability could begin from the international economic and political framework and derive the paths that national economies should follow to reach and maintain global equilibrium. Such a deductive approach is far from realization today. My purposes here are far more modest: to consider another example of institutional success, to compare it with the Queensland and Columbia experiences, and to highlight the vagaries of social learning in the diverse human community of international organizations.

Pollution in the Mediterranean. A nearly closed body of water bounded by industrial nations and developing countries struggling toward economic advancement, the Mediterranean Sea faces serious threats from pollution of its waters. This is an irreducibly international problem: pollutants from one nation foul the coasts of others. The responses of the nations whose coastlines touch the Mediterranean were examined in 1990 in an influential study by the political scientist Peter M. Haas. Pollution, Haas observed, "challenges the core of the international legal order," which is founded on the proposition that nations exercise sovereign control over their own territories. The result is a tragedy of the commons: "Different countries wished to use the Mediterranean for different purposes, and hence disagreed about what level of water quality was desirable. Moreover, leaders disagreed about the need for immediate action, the range of pollutants to control, and the fundamental desirability of negotiating with countries that had traditionally been enemies. With a large proportion of their industry and population lying near the coast, many [developing countries] interpreted efforts to control marine pollution as indirect ways of retarding their attempts at industrialization." In this distinctly unpromising situation there was no incentive to cooperate, and no way to enforce cooperation even if it were agreed to.

What happened was a surprise: a broadly drawn Mediterranean Action Plan (Med Plan) adopted in 1975 by 16 nations, later expanding to 18 plus the European Community. In accordance with the Med Plan, national pollution-control policies were enacted into law and implemented to some degree in the years after 1975. Although the pace and scope of compliance have been far from uniform, any enduring form of cooperation in the Mediterranean basin is noteworthy.

Epistemic community. What made the difference was the political role of experts. Knowledge of environmental problems, when organized appropriately, exerts political influence. Haas labeled the form of organization that proved effective an "epistemic community."

An epistemic community is not a formal organization but something similar to the advocacy coalition discussed in Chapter 4: a network of experts who share a belief in the importance of pollution as a problem and a set of priorities for dealing with it. The epistemic community has an additional property—"a common approach to understanding." That is, what binds the community together is not only substantive interest but also agreement on procedures for processing information. "Unlike an interest group, confronted with anomalous data [an epistemic community] would retract their advice and suspend judgment." An epistemic community is thus similar to a community of scientists who, with their aversion to Type I errors, would reserve or withdraw judgment when faced with facts that do not fit their established understanding.

An epistemic community such as the scientists, lawyers, and other experts in the maritime pollution of the Mediterranean becomes important in politics when the community provides insight into important problems.[2] Faced with an epidemic, politicians listen to experts in public health and medicine, imposing controls on travel or other activities because the experts' recommendations can solve the political problems caused by rampant disease.

Turning an academic network into a political instrument required organization. Haas traced how the fledgling United Nations Environment Programme (UNEP), created at the 1972 Stockholm conference on the human environment, took on this organizational role: "UNEP's broad viewpoint enabled it to form transnational alliances with marine scientists and nongovernmental organizations. . . . These alliances were cemented with the provision of research funding, monitoring equipment, and training in its use. Together they led governments to accept a broader international agenda and to support more comprehensive domestic policies." Whether they were already advising their governments, were working in government agencies, or were not directly connected, scientists came to share an understanding of the problem of pollution of the Mediterranean, "conveyed congruent advice to their governments, and policies began to converge." The credibility of their advice lay in the fact that their understanding was scientific—that is, open to revision by new information. At the same time, UNEP could harness scientists' interests by sponsoring research. The result was to organize a politically effective community of credible actors. This led to changes in national policy and learning: "government learning came from the political power of the epistemic community, once it was entrenched in domestic administrations. . . . The epistemic community usurped control over environmental policymaking, and shifted policy in accordance with their shared values and understandings. Regional scientists learned of new techniques and data from each other, through the networks sponsored by UNEP." This is an advocacy coalition with influence.

Influence is not control of policy, however. Although UNEP urged regionally integrated planning, so that pollution control could be monitored and adjusted throughout the sea, the national governments preferred to proceed autonomously. A coherent data base for the Mediterranean basin is a precondition for thinking about the sea as an ecosystem. But there is scant support for regionally consistent monitoring, since such information would identify those nations that pollute more than others. "The form of learning was that which was least threatening . . . to states' pursuit of autonomy and security."

UNEP's policy-driven interest in science was catalytic but ambiguous in the long run. In recent years, according to Michael Scoullos, a marine geologist and environmental activist, UNEP has continued to press the cause of regional planning, but it has gradually lost the support of the scientific community in this effort. Building a data base for planning does not mesh with current enthusiasms in marine science, and European scientists have drifted off to other funding sources in the European Community and their national governments. Scoullos does credit UNEP with playing a central role with scientists from south of the sea, however, for whom the inducements of new equipment and training were valuable.[3] Having better scientific expertise, in turn, facilitated their participation in the international arena.

In sum, the epistemic community was a way to get around some of the barriers to joint action erected by national interest, but whether the effects will endure—or even widen to an ecosystem perspective on the Mediterranean—is unclear. Without an ecosystem perception of the Mediterranean Sea, there is of course no possibility of experimental learning at that scale.

Multilateral institutions. Is the Med Plan fostered by an epistemic community a harbinger or a curiosity? Ernst Haas, a distinguished theorist of international relations, provides useful context for his son's findings, together with a lesson about learning in institutions: learning is unusual, and is typically driven by a crisis that is used by a prepared leadership. According to the elder Haas, international organizations such as the United Nations or the Med Plan are creatures of their members: "The clients are at the same time the masters and paymasters. The staff and management serve at the pleasure of the clients. . . . Consumers can vote the management out of office. International organizations exist only because of demands emanating from the [institutional] environment and survive only because they manage to please the forces there." This condition, which Haas calls "hyperdependency," is important to learning in large ecosystems because the coordinating planners responsible for experimental learning are likely to be highly vulnerable in this sense. Large ecosystems lie by definition in the jurisdiction of more than one government; political pressure is endemic.

International organizations are formed to deal with problems by nations that would prefer to remain completely sovereign but find the costs of doing so too great. The problems that arise are often "motivated errors"—situations like the tragedy of the commons, in which what is wrong is inherent in the way parties interact with one another. In these cases, the fact that the parties are *being* rational is a central part of the problem.

Crisis and learning. Hyperdependency and motivated errors frequently put international organizations in crisis. According to Ernst Haas, crisis is an opportunity to learn: " 'An international organization learns' is a shorthand way to say that the actors representing states and members of the secretariat . . . have agreed on a new way of conceptualizing the problem. That is, it is not individuals, entire governments, blocs of governments, or entire organizations that learn; it is clusters of bureaucratic units within governments and organizations."

This definition recalls the idea of the advocacy coalition. Learning is a change in the understanding of a network of people, whose members are located in different organizations. The change in understanding is instrumental—a new way to conceptualize a problem. But usually the solution to the reconceptualized problem will require reconfiguring the links between institutions. Those rearrangements require power, and often lie beyond the reach of those who have learned. The change needed to achieve a solution, Haas concludes, "requires a crisis, a shared definition of its causes, and organizational leadership willing and able to profit from the combination." The instrument of that leadership is an epistemic community, which is called upon to formulate a different conception of the problem to be solved—that is, to propose an altered mission for the international organization. If the change is successful, the international organization is redirected and its members have a different understanding of its mission and of how their interests are linked.

Ernst Haas summarizes the responses of the World Health Organization, UNEP, the World Bank, and the International Monetary Fund, showing how learning in the sense of reconceptualizing the problem to be solved took place. He also surveys failures. In almost all these cases, positive and negative, international organizations were endeavoring to solve a major problem of developing countries, on behalf of the industrialized nations. Haas cautions, however, that there is a large difference between learning in an international organization and learning in the target populations it is intended to help: "the role of the grassroots is very remote." This difference may also apply to international efforts to move toward sustainability. Learning is not simply an intellectual process: "Learning is not sudden enlightenment or even incremental insight. It is the establishment of shared meanings among parties that may be active antagonists but that find themselves condemned by their interdependence to negotiate better solutions than they had created in earlier attempts."

The "establishment of shared meanings" is a more elegant way of describing the negotiating process described in Chapter 4. Shared meanings can promote better management of interdependence. Moreover, this kind of learning can both synthesize the problem to be worked on in a new vision, and analyze it into tasks feasible for human organizations to carry out. As I pointed out in Chapter 4, "planning centralizes, implementation decentralizes." Learning to manage interdependence has yet to be institutionalized anywhere, however, because crisis appears to be a necessary condition for transferring the reconceptualization within its original network of professionals into institutional arrangements of some permanence.

Haas's studies of international organizations put learning at large scales into a different light: learning changes the learners' understanding of what they are trying to accomplish together. What is learned has as much to do with goals as with means. In this process, an epistemic community plays a strategic role in developing alternative ways of understanding what *can* be done within the limited capabilities of a "hyperdependent" organization. The international organization's mission plays the role of the hypothesis in adaptive management: the mission becomes a premise to be tested by action, in the expectation that it will often be wrong.

Being wrong creates crisis, an occasion for further change. Because crisis is chaotic, learning is not institutionalized. But learning *does* produce something essential for sustainable development: a shared understanding of what can be done. It is a provisional understanding; what matters is not that the right answer emerge—that is a utopian hope. What matters is that better questions can be posed. "Moral progress," Haas argues, "must be defined in procedural terms."

The cases sketched above are neither representative nor comprehensive. It is suggestive, however, that none has gone further than the Columbia basin program in linking human interventions to learning about nature. In each of these cases, knowledge gained by experiment would improve either the goals pursued or the means by which they are achieved. In each of these cases, the capacity to learn is problematic but potentially transforming when it occurs. And even in the "hyperdependent" international organization, it appears to be possible to survive the conflict and turbulence of crises.

These are better results and clearer warnings than we might have expected. Learning is possible and worthwhile, but it is difficult and intertwined with crisis and conflict. That conclusion, grounded in experience, suggests the importance of expectations: what sort of learning can we anticipate in the world as we find it? For this question, theory is a richer vein to work.

Chapter 6

Navigational Lore:
Expectations of Learning

[Learning is the] reconstruction or reorganization of
experience which adds to the meaning of experience,
and which increases ability to direct the course of
subsequent experience.

—John Dewey, *Democracy and Education*

At the end of Chapter 3 I defined a task: to find practical learning
strategies for large ecosystems. The comparison cases of Chapter 5
illustrate what has been attainable. Here I link these experiences with
theories of learning, to make explicit the expectations we should
bring to learning at the ecosystem scale and over times of biological
significance.

I have organized theories describing learning along two dimensions: by the assumptions each makes about the character of the
learner, and by the assumptions each makes about the way the
learner makes decisions. I do not claim that any one theory is best.
Managing large ecosystems is a rudimentary art, and no single theory of learning is likely to be usable or helpful in all cases.

Learning involves different things for an individual than for
groups, and different things for groups made up of people who have
multiple, perhaps conflicting objectives than for groups sharing a
single set of purposes. When there is a single set of purposes, the
learner is a *purposive* entity, a genuine organization. If the members of
a group do not share purposes, the learner is a *collective* entity, with an
everyone-for-himself potential for anarchy. Theories about individual learning are better developed than those about group learning,
but they seem less likely to be realistically applicable to the multiple
and contending interests of a large ecosystem.

Learning from experience occurs when decisions produce results.

Learner

	Individual	Purposive	Collective
Rational choice	Rational	Double-loop learning	Principal versus agent
Bounded rationality	Cybernetic _Superstitious_	Single-loop learning	Coalition-bound incrementalism
Biased cognition	Heuristic	Institutional constraints	Organized anarchy

Decision process (label at left, spanning rows)

FIGURE 6–1. Some theories of learning.

The conventional assumption about decision making is that it is based on rational choice: the decision maker examines all the available options and selects the one that produces the best result. In complex situations, however, such rationality seldom prevails. Decisions tend either to be based on limited consideration of a limited number of options—the situation called bounded rationality, sketched at the beginning of Chapter 3—or to be skewed by biases in cognition, limitations in human judgment that are similar to optical illusions. These distinctions produce an array of nine theories of learning, shown in Figure 6–1. Each represents a set of assumptions that leads to predictions about how humans behave under specific circumstances.

Rational Learning

As the cost of electricity from coal and nuclear plants rose further and further above the cost of hydropower from the Columbia's dams, energy conservation became increasingly attractive to utilities as an alternative to increased generating capacity. As experience showed that energy efficiency could be developed at low cost, conservation displaced generation as a priority in the Northwest. This is a story of rational learning: an individual actor ("the Northwest") updates understanding and modifies choices as new information improves comprehension.

Most of the time an actor has more than one goal, and the goals are partially incompatible: development conflicts with sustainability, fairness with economic efficiency, and so forth. In the rational model, the competing goals are viewed all together and a resolution achieved: tradeoffs, compromises, and creative solutions integrate conflicting values into a coherent set of goals. In the Pacific Northwest power system, the rule of cost-effectiveness governs integration. As conservation is developed, the cost of the remaining projects rises, and research lowers the costs of technology. When the cost of generating electricity equals the cost of the remaining conservation efforts—including estimates of all social and environmental impacts, such as damage to air quality from coal-fired power plants—the least expensive generating alternative is developed next. This framework of least-cost planning shapes the Northwest Power Planning Council's energy plan.

Yet rationality is the exception rather than the rule in human life, particularly in politics. The rational vision is prescriptive: it says what we believe should be done, and that matters even if—often, precisely when—what is done is not rational. The influence of a rational model of decision making goes beyond defining a standard against which to measure shortfalls, however. Rational *learning* is assumed in most policies. Thus a recent discussion of policies to manage risky technologies included the following recommendations:

- *Protect against foreseeable serious hazards.* For example, the National Environmental Policy Act, requiring examination of the environmental impact of major actions, warns of potential harm on the basis of past experience.

- *Err on the side of caution.* For example, the Endangered Species Act, aimed at recognizing and taking special measures for species under pressure from human activities, relies on past experience to reach judgments of what populations are endangered.

- *Experiment to reduce uncertainty.* For example, the testing of pesticides guards against harms by requiring laboratory or pilot-scale experimentation. (This is rarely ecosystem-scale learning of the kind described in Chapter 3, which could sense indirect effects such as harm to bird life, whereas laboratory-scale testing can find such effects only if the experimenter knows to test specifically for it.)

- *Learn from experience.* For example, initial fears about release of life forms created by biotechnology led to careful monitoring of laboratories using these methods, until it became accepted within the technical community that the risks were in fact not very high. This sort of learning is infrequent: there has been little use of environmental impact statements, which are supposed to predict impacts of large projects, to see if the projects do in fact produce the expected impacts on the environment.

• *Focus resources on the most important hazards.* So long as resources are scarce, it is vital to use those available in the most efficient way. Alarmingly, this was, until very recently, a revolutionary idea. The battle over nuclear power demonstrates, however, that the definition of "important" hazards can be highly controversial.

These sensible ideas rely on rational learning. They imply that managers can be held to account for learning, and that reforms should aim at overcoming barriers to rationality. But in fact information that comes from experience is often put to other than rational uses.

Principal versus Agent

In any organized setting there is an inevitable divergence of interest among the actors: doctors have different priorities from their patients, bosses from their subordinates. Economists call this the "principal-agent problem." Principal-agent theory highlights the role of information in the structuring of human relations, and it demonstrates how individual rationality and learning can lead to perverse results—individuals may or may not do better, but the population as a whole can end up worse off. Individuals' abilities to pursue their interests may decrease overall performance and thwart institutional learning. Rational learning among individuals can diminish collective outcomes.

Consider fish hatcheries. Hatcheries are an alternative to saving or rehabilitating habitat in order to produce salmon. They require continuing operation, budgets, and staffing. Prized by fish bureaucracies, hatcheries are fiefdoms in fish and game agencies. Virtually all these agencies in the Columbia basin—self-proclaimed stewards of nature on behalf of society—continued to be ardent defenders of salmon hatcheries long after scientific information began to accumulate on their biological and ecological problems.

This phenomenon is widespread. Most individuals cannot achieve their objectives without someone else's help. The person or organization seeking to get something done is called a "principal," and the person or organization providing the means to do it is called an "agent." Sometimes both principals and agents are individuals—client and therapist, hunter and guide—but frequently agents are organizations, such as governments.

According to principal-agent theory, organizations are necessary in the economy because information flow is constrained: principals often do not know or cannot evaluate what agents are doing. The principal knows better than the agent what he or she wants, but the agent knows better how to get it and the cost of doing so. The agent is often in a position to shade the truth about costs, to guide the principal to inferior results, and generally to serve the agent's interests at the expense of the principal's. This temptation results from the way information is distributed. Organizations encourage or require agents to serve principals by creating incentives, providing information, and facilitating enforcement. Thus institutional arrangements and organizations may be understood as solutions to the problems of agency.

Markets can protect principals in their role as buyers. Competition ensures that price and often quality bear a fair relationship to the cost of providing a good or service. The problems start when competition is weak or absent. When information on the performance of an agent is hard to obtain, or when agents have objectives opposed to those of principals, there is likely to be trouble. Such relationships between principals and agents are typical in the public sector, where services such as police protection or wildlife management are difficult to monitor and where taxpayers' willingness to pay is usually tempered by the suspicion of malingering on the part of government employees and their political masters.[1] When there are long delays, technical complexities, and fractiousness of the kind found in large ecosystems, the problems of agency are endemic. Yet as the Columbia basin experience implies, creating a comprehensive management organization to overcome the problems of agency is often politically infeasible.

It is accordingly useful to consider how economic organizations have evolved mitigations to deal with agency problems:[2]

- Although market prices may not be available, *monitoring of an indicator* of performance is often possible. When monitoring concentrates on inputs rather than on results, however, there may be perverse incentives. As salmon populations have declined, the typical response has been to increase artificial production, even though doing so further weakens natural stocks of fish.

- If the agent's prosperity is tied to his or her *reputation*, or if the agent relies on assets such as manufacturing facilities subject to litigation, it will be rational for an agent to serve the principal's interest whenever the gains from any particular transaction are lower than what the agent stands to lose. This principle is a widespread deterrent to misbehavior.[3]

- *Long-term relationships* build up stocks of reputation and expectations usable if enforcement should be necessary, and they make limited monitoring more effective. From this perspective a benefit of formal organization is that it fosters durable relationships. The emergence of working relationships among utilities, Indian tribes, state fisheries managers, and nongovernmental organizations has been an essential condition for coherent management of the Columbia basin ecosystem.

- When there is some competition, the benefits and *losses* from agency problems *are shared* by both principal and agent; the used-car dealer who cannot sell good vehicles for what they are worth suffers, just as does the used-car buyer who must worry about the quality of her or his purchase. Therefore, principal and agent share an incentive to structure their relationship to produce outcomes as if information were free.[4]

- Some incentive structures entail pure losses, such as imprisonment for criminal activity. In these cases in which there is no way to collect appropriate compensation the only deterrent is to impose a *penalty on the transgressor* without compensating benefit to the victim.

Methods of monitoring and enforcement are responses learned over long periods, ways of dealing with the risks inherent in agency by assigning institutional roles and handling information through accepted routines. These methods are grounded in precedent, rather than being motivated by the desire to prevent error or evil in the future. The mitigations can therefore be functional (they can work) without being rational (grounded in reason). As a form of learning, such traditions share the frailty of lessons that have been detached from the experience that taught them. More generally, as circumstances change, it is unclear when tradition should give way to experimentation because there may be no lingering memory of why the tradition was originally adopted.

Principal-agent theory is instructive about several aspects of institutional learning:

- Even when actors are individually learning in a rational fashion, the outcomes they jointly produce may be inefficient because the actors' interests are divided.

- The tension between principal and agent arises from the cost of gathering and transferring information. The harder it is to monitor the activity of the agent, or the more costly such monitoring becomes, the more likely it is that there will be mistrust, malfeasance, or other sorts of trouble.

- There are adaptations that mitigate or control these effects in many instances. They tend, however, to be evolved behaviors that are used without an understanding of their functional dynamics. In the face of changing or unfamiliar circumstances, it may be possible to devise appropriate adjustments by designing incentives for agents and principals so as to lower the cost of information transfer.

Each of these problems is prominent in large ecosystems, where the roles of principal and agent are shared higgledy-piggledy among the entities exercising partial control over pieces of the ecological domain. Chapter 4 described means to respond to the splintered interests of large ecosystems; principal-agent theory gives a simple (though incomplete) way of thinking about those means.

The incentives that face principal and agent are inherently different, and the operation of these incentives divides their interests. These characteristics are signs of bounded rationality: though agent and principal share some objectives, each must pursue a different path to reach them, and there can be conflicts of interest along the way.

Bounded Rationality

Much of what we mean by "learning" comes from thinking about individuals, but the entities that need to learn about ecosystems are organizations and networks of organized interests. Learning how to do better may be one of the objectives of formal organizations such as the Bonneville Power Administration, but it is not the principal objective. Institutional networks, such as the one composed of the Indian tribes, electric utilities, and government officials who wrestle with allocating the flow of the Columbia, do not have overall purposes that enable the network as a whole to judge outcomes and learn from them. Individuals and organizations do learn from their interactions with each other and with nature, but they seldom do so deliberately, and what is learned has more to do with competing against one's rivals than with direct benefits to fish or ratepayers.

At a different level, it is not clear what the relationship is between learning by individuals and change in institutional behavior. Institutional veterans understand much more about a situation than they put "on the record"—indeed, knowing what can and cannot be said publicly is part of what qualifies them as seasoned veterans. So what an individual learns may not be put into the retrievable memory of an organization. Conversely, changes in institutional behavior, such as the assertion of Indian treaty rights in fishing, may or may not be accompanied by individual learning or changes in individual attitudes.

The principal tool for untangling individual behavior and changes in organized human relationships is the idea of bounded rationality elaborated by Herbert A. Simon. Humans are *not* rational in the idealized sense of considering all alternatives and selecting the best course of action at each time. Instead, people employ *bounded* rationality: the limited consideration of a limited number of options in making a sequence of decisions. Some implications follow:

- Nearly all decisions aim at *satisfactory* rather than optimal outcomes. That is, doing one's best does not lead to the best results, simply because in most circumstances the individuals involved are unable to comprehend or identify the optimum choice.

- The implicit contract between manager and worker (or, more generally, between principal and agent) defines a *zone of acceptance*. That is, there is a range of activities in which the worker permits the manager to define the goals to be met by the worker. In turn, the manager accepts satisfactory rather than optimal performance from the subordinate. The zone of acceptance may be marked off in many dimensions—the time and resources allowed must be feasible, the goal must be consistent with ethical or professional norms, and so forth. These zones of acceptance produce a division of labor and responsibility. When successful, this division creates a human organization that can achieve, through teamwork, objectives that no individual member could reach, but in a way that does not require superhuman performance from any single member.

- Bounded rationality is not guaranteed to work; any organization can fail to solve specific problems. In particular, because consideration of problems is limited, some conflicts among goals will be ignored, and some goals will be ignored at least temporarily.

- Rapidly changing environments pose particularly difficult situations for bounded rationality. The bounded decision maker requires problems to conform to an established framework, if they are to be recognized and solved within existing zones of acceptance.

Bounded rationality is *procedurally rational*: a general formula for dealing with complex tasks and complex environments, by focusing the attention of individual members of an organization on narrow problems they can handle within their zones of acceptance. To the caricature of the functionary in a large institution—one who knows his job, but only his job—Simon added the realization that narrow expertise is a necessary consequence of the way complex purposes are realized by human organizations. Procedural rationality, in turn, harbors a surprising implication for deliberate efforts to learn: the decision to invest in learning tends to vary with the success of the organization, not with whether learning is effective.[5]

Cybernetic Learning

The procedural emphasis of bounded rationality in organizations produces an important result called *cybernetic* learning: change in behavior that does not integrate information, permitting incompatible goals to remain detached. Facing a complex world, the decision maker seeks not the best outcome among competing objectives but instead a satisfactory result on each goal, taking the goals *one at a time*. Thus, instead of solving a set of linked problems within a coherent strategy, the decision maker monitors a few critical variables and tries the keep the system within the bounds defined by limiting values of those variables. The cybernetic decision maker is like a driver in traffic, steering to avoid other cars, making turns and speed adjustments, but not attending to the overall flow of a large number of cars during the rush hour. As this analogy suggests, the cybernetic approach can work for the individual but may not be optimal or even functional for the collectivity. The cybernetic organization learns simply what works, without having to understand what works best or even why its own behavior is satisfactory.

Utility managers in the Northwest projected rising consumption of power and acted to supply it, ignoring the fact that the price of new generation was so high that rates would skyrocket. They said, when asked, that electricity was a necessity of life, that people cared more about an adequate power supply than about its cost. These had been critical variables in a regulated monopoly whose focus for decades had been on technology and finance rather than on economics and customer relations. But the price rises, when they came, coincided with a deep recession and forced the families of jobless workers to choose between heating and eating. The ensuing uproar drastically enlarged and redefined the critical variables utility managers had to keep within bounds thereafter.

Cybernetic learning exemplifies the caricature of "the left hand not knowing what the right hand has done," as different parts of a large institution learn from their localized and bounded experience in ways that may cause them to work at cross-purposes. In fisheries management, the success of a hatchery may be undermined by inappropriate regulation of harvest. The fishing fleet, for example, may intercept the adult fish needed to spawn the next generation. Part of what makes learning from experience problematic, then, is that what counts as experience and what is to be learned from it can vary with one's position within the institutional structure that enacts the policy. A dictum of political science, "where you stand depends on where you sit," summarizes working wisdom about conflict. It affects learning as well.

The cybernetic model conceives of decision making as the monitoring of a small number of variables, with responses designed to keep these variables within acceptable limits. This approach assumes either that the environment of the institution is stable or that changes in the environment are so clear-cut that a small repertoire of managerial responses will suffice to keep critical variables in bounds. It does not matter whether the decision maker understands either the causal factors that drive the critical variables or the significance of these critical variables in a larger scheme of goals. Competing values are therefore not integrated or traded off against one another. That is the bad news. The good news is a characteristic of all types of bounded rationality: people in organizations can do things jointly that they do not comprehend individually. That is how it is possible to build an electric power system or operate a hospital or accomplish other things that are literally beyond the capability of any single person to understand.[6]

An important instance of cybernetic decision making is the development of a budget within a large organization. As a way of managing complexity, overall objectives are separated into goals to be met by separate subunits, using the resources allocated in the budget. Where there are linkages that couple a low-priority objective in one unit of the organization to a high-priority one somewhere else, the separation of goals can be dysfunctional.[7] Environmental quality frequently cuts across functional lines in an organization because environmental effects may be of concern at different stages of economic production or organizational action. At the same time, environmental concerns are typically of low priority in some parts of the organization; for example, those charged with production are likely to be less concerned about the pollutants they generate than those responsible for relations with regulatory agencies. It is therefore not enough to declare an overall commitment to environmental quality.

If the commitment is to have substance, environmental activities need also to be included in budgets and assignments of personnel; such commitments of resources, in turn, can lead to conflict among units of the organization as environmental activities and those who carry them out interfere with established routines.[8]

An extreme case of cybernetic learning sheds light on why idealism about science is difficult when it is most needed. Learning is *superstitious* when an inferred cause has no actual role in bringing about its supposed effect. Consider the behaviors of sport fishermen who perform idiosyncratic rituals in attempts to improve their success, or athletes who wear a "lucky" item of clothing while on a winning streak. Psychological research has unsettling suggestions about the question of whether the causal explanations we believe may in fact reflect random associations. Superstitious learning is especially likely when objectives are affected by performance—for example, in a competitive setting, in which one's objective is to do well in relation to others. Under such conditions, whether one meets one's objective or not is likely to be random, and explanations of how one changes one's performance are, ironically, both psychologically compelling and likely to be wrong.

Attempts to rescue environments that have been severely damaged typically create situations in which objectives are affected strongly by performance, and thus in which incorrect but intensely held explanations are likely to gain credence. The fierce debate over how to modify the flow of the Columbia to benefit migrating fish continues to be conducted with few facts, as we saw in Chapter 2; it is simply too early in the learning process to know if the water budget and other more costly measures will work to reverse high mortalities in migrating juvenile salmon. This ignorance raises rather than lowers the heat of the arguments, in which divergent values are reinforced by causal beliefs of dubious validity. Statistically reliable learning may come too late to save imperiled fish. It certainly will come only after large expenditures of funds paid by electric ratepayers. The battle over what to do in the short term can easily undermine the longer-term need for learning. In this way, conflict reinforces the tendency toward superstition.

As we saw in Chapter 3 in the discussion of regression toward the mean, both the behavior of an endangered population and the inferences humans make about what is causing its distress are likely to be confounding. In such situations the rigor of experimental skepticism—that is, adaptive management—is especially valuable. Experimentation will also be controversial, however, because the suspension of belief necessary to test hypotheses scientifically must persist in the face of fervent superstition.

Superstitious learning is a limiting case of cybernetic learning. The learner seeks to maintain critical variables by doing things actually unrelated to achieving the objective. Superstitious learning is common when objectives are related to outcomes. In those cases especially a skeptical, scientific approach is both useful and difficult to sustain.

Coalition-Bound Incrementalism

Division of power within an institution results in a collective form of cybernetic behavior. The way in which large organizations make choices is constrained by incomplete agreement among influential persons. In contrast to the neat hierarchy of the organization chart, social scientists observe that government agencies and corporations are governed by coalitions of their leading players. Many significant choices strain the coalition; for example, declining sales, a problem for the marketing staff, may increase pressure on research to develop new products, or on production to lower costs. The solutions adopted under these conditions are typically constrained by the need to maintain the coalition, since control of the whole organization requires coalition members to work together. (That is what defines membership in the leadership coalition: one is indispensable because one represents a functionally necessary group.) Solutions that are acceptable to the leadership are usually similar to solutions used in the past because such responses leave questions of power and privilege undisturbed. Precedent shapes the search for answers, and the answers tend to be incremental.

Decision making by coalition is necessary when there is no overall organization with a clear purpose, but cooperation is necessary to manage a shared resource. Water in the Columbia River poses just this kind of challenge; therefore, it is not surprising that changes in river operations to benefit migrating salmon have been both contentious and slow to adapt to the changing conditions of fish populations.

Learning is cybernetic in a coalition because the members of the coalition are unlikely to want to align their divergent goals into a single coherent set of objectives. That is why cooperation takes the form of a coalition in the first place. Research by Richard M. Cyert and James G. March suggests that coalitions will also tend to learn by incremental changes rather than by profound ones because the former are more likely to be acceptable to coalition members.

Single-Loop versus Double-Loop Learning

An organization can be thought of as a social technology designed to perform a specific set of tasks of production and distribution. The organization is thus a working model of a theory for solving a specific set of problems. Usually that theory does not perfectly represent the tasks facing the organizations. Reality is more complex than imagined by those who act within it. Errors therefore emerge. Some of these can be solved by mechanisms embodied in the organization because they are already anticipated in the underlying theory on which the organization is based. This process has been labeled *single-loop learning*.[9]

Other problems are not recognized or not soluble by the theory embodied in the organization. Overcoming such problems, a process called *double-loop learning*, requires rethinking the purposes and rules of operation so as to diagnose the problems of theory that underlie practical problems. Searching self-examination of this kind is, understandably, a difficult task for all members of the organization. Consequently, in many cases double-loop learning is not even attempted. Even the need to reexamine purposes is often invisible because, to those with only a partial grasp of either, the mismatch between theory and reality usually looks like selfishness, stupidity, or evil on the part of others in the organization.

Once state fisheries managers understood that making common cause with the Indian tribes could greatly increase the resources devoted to salmon enhancement, the old hostilities began to abate. A new, joint purpose emerged, around which cooperation made sense. Both tribal and non-Indian fisheries managers adopted the language of ecosystem management—a different way to understand what they were trying to do. At the outset much of this adjustment was superficial, but as joint projects were carried out, new relationships among individuals developed. Not all of these are free of bias, and some involve enmity more than cooperation. But the texture of dealings among organizations changed with the individual relationships. If ecosystem management is a better way to nurture the salmon fishery (in both its human and nonhuman dimensions), then real learning will have occurred. Although it is too early to conclude that the ecosystem perspective *is* better in the Columbia basin, it is clear that hard-to-reverse institutional changes have taken place.

Cybernetic and single-loop learning can be thought of as complementary in their failings. Cybernetic learning largely involves failure, at the lower levels of an organization, to appreciate the structure and interdependence of goals higher in the hierarchy. Double-loop learning involves failure to learn from subordinates' experience: knowledge gained at the working level is not reflected in higher-level decisions, because doing so acknowledges defects in the operating theory of the organization. That is why crisis is both recurrent and necessary in the learning process of international organizations, as discussed in Chapter 5.

Adaptive management can be understood as a way to incorporate double-loop learning into the routines of an organization, giving double-loop learning some of the strengths of bounded rationality. The experimental approach adopts a skeptical and tentative view of the theories humans use when attempting to manage a natural system. Those theories are tested against experience when management is adaptive in design and execution. Theories should then be revised in light of experience. The discussion of double-loop learning suggests that this rationalist picture does not put enough emphasis on the institutional conflict that is likely to ensue. As organizations change their theories of action, managers' control is likely to be undermined because the reasons for exercising control are being challenged and revised. That is why conflict is an inherent part of social learning, a theme developed in Chapter 4.

Yet if conflict does not tear the institutional fabric, the result of double-loop learning is an improved theory of reality—that is, real learning, in which the operating premises of the institution are brought into closer correspondence with actual responses, and in which the cybernetic disconnection of objectives can be realigned into a coherent whole. Most important from a practical standpoint, routine problem solving can then resume in a single-loop fashion, handled effectively by the more accurate theory of reality. When real learning is essential, the risks of double-loop learning may be part of the price to be paid.

Biased Cognition

The existence and effects of superstitious learning suggest that psychological phenomena matter. Perhaps because our species evolved in a world different from the one we live in now, there are systematic errors in human perception. As in the case of the cybernetic separation of goals in organizations, individual and collective human perceptions can be disconnected from reality under some circumstances. These are not pathological conditions in the sense of mental illness, but are analogous to optical illusions—situations in which the inferences people draw are systematically mistaken. Important and long-lasting distortions of learning result.

There are four major kinds of cognitive bias, which can exist not only in the perception of individuals but also in the ways in which organizations attend to problems. First, there are biases that arise from an incorrect inference of the relationship between cause and effect—superstitious learning, discussed above as a limiting case of cybernetic learning. Second, there are biases of perceiving evidence, such as regression toward the mean already discussed in Chapter 3, and heuristics or shortcuts used to draw inferences, discussed below. Third, there are biases induced by social interactions, such as organized anarchy, mentioned briefly in Chapter 4 and discussed at greater length here. Finally, there are biases that affect organizations rather than individuals, arising from the way attention is allocated in accord with organizational purposes. These I call institutional constraints.

Common to all four kinds of bias is skewed reasoning, either on the part of individuals or in organizational processes. These systematic errors misdirect learning: surprises can be overlooked or rationalized; the wrong lessons can be learned; and misunderstanding and conflict can result even when objectives or interests are not really mutually exclusive. Cognitive bias is an intrinsic property of human information processing. It cannot be eliminated at the individual level. Instead, where biases are costly and predictable, institutional mechanisms can be designed to recognize and correct mistakes in information processing. This correction is not free, and the correction process itself is subject to error and bias. Understanding these systematic effects is indispensable if one is to recognize, avoid, and cope with them.

Heuristic Biases

Ordinary life is sometimes not a good model of reality. This unsettling conclusion of psychology can be seen in unhealthy behaviors such as smoking—the long delay before its health effects are felt masks the consequences from smokers. People use shortcuts in judging the likelihood of uncertain events. As in the case of optical illusions or bounded rationality, simplifications of reality sometimes lead to erroneous inferences.

Two biases that mirror each other are of special importance in governing large ecosystems. These are the systematic tendency to overestimate the likelihood that plans will succeed and a corresponding tendency to underestimate the probability that something will go wrong in a complex undertaking. Both involve a failure to appreciate the quantitative impact of many events that are linked to one another. If there is any chance that *some* of these events will go awry, the fact that they are all tied together greatly reduces the chance of success of the whole. For example, the significance of building a sequence of dams along the Columbia was underestimated. Although each dam kills a relatively small fraction of the juvenile salmon migrating through it, their combined impact is much greater, since the fish that survive passage through one dam must still run the gauntlet of the others. The result is a disastrously low survival rate.

Perhaps more worrying is the "law of small numbers." This is the tendency to think that small statistical samples will have the properties of the larger universe from which they are drawn—the expectation, for example, that in tossing a coin four times one is very likely to get two heads and two tails. (The actual probability is three out of eight. One should not bet even odds of getting two heads and two tails, since the likelihood of *not* getting two heads and two tails is five out of eight, or 62.5 percent.) People make unreasonable generalizations from a few cases, and in uncertain situations they think they are wrong more often than they should.

Consider the question of whether fish hatcheries work. Hatcheries are expensive, so not many are built. Moreover, the first hatcheries to be built are likely to be imperfect because the designers and operators are still learning. But if the first few facilities are disappointing, the law of small numbers will lead people to conclude that hatcheries are a failure, period. This is the phenomenon of Type II error discussed in Chapter 3: believing a proposition (hatcheries work) to be false when it may be true.

Of course, we don't know if hatcheries *do* work. The right way to find out is to use statistical methods to judge the probability of a Type II error. But that undertaking requires a precise statement of the proposition to be tested and a systematic, experimental approach to gathering data. In short, it requires adaptive management. Without careful testing, all we know is that people are very likely to rely on data that are insufficient to support the conclusions they hold. There are few ways to rehabilitate damaged ecosystems; it would be tragic as well as foolish to discard one that actually had promise because of a statistical illusion. It is important to resist the misleading persuasiveness of small numbers.

Another striking bias involves the tendency *not* to simplify reality enough, which psychologists call the "relevance of irrelevant information." Although people can make unbiased judgments when they are given only bare quantitative information, those judgments are readily displaced when subjects are given information that has no substantive bearing on the estimates they are asked to make. For example, if we knew that habitat restoration projects produced measurably more salmon in half the cases, while egg boxes (devices that protect newly hatched salmon) worked one-third of the time, that information could be used to set up a program to rebuild a population while controlling costs. But if a policymaker is *also* told that the habitat restoration sites are mostly on Indian reservations, while the egg-box sites are in state-owned land, the additional information—which is not apparently relevant—may well produce a different perception of what is at stake, and a different choice. In a world in which most decisions to invest in environmental protection or natural resource management turn on information that bears on a range of dimensions, it is hardly surprising that choices often have little to do with biological success.

Errors in decision making that come from the heuristics used in everyday life can be corrected by systematic methods, of which the best developed is statistical analysis. Unfortunately, policymakers rarely grasp the implications of Type II errors in statistical interpretation, and so far their scientific staffs have been slow to instruct them on the need to guard against these errors by formal analysis.[10]

Institutional Constraints

While superstitious learning and cognitive biases in individuals reflect misunderstanding at the individual level, error can also take the form of inappropriate social organization. Western water law in the United States is a deeply rooted social institution that fosters misunderstandings, forces those who would deal with the problems to act in dysfunctional ways, and skews the lessons taught by experience away from understanding. Water in the West belongs to individual users, whose property rights in recent years have been subject to water-pollution regulations. Much else lies wholly under the user's jurisdiction. Farmers in the tributaries of the Columbia have no obligation to leave enough water in the streams for salmon to hatch or grow or migrate, and losses at all three stages have diminished naturally spawned populations. The water right legitimates the benefits belonging to the holder of the right; it does not acknowledge others' claims, even though flowing water is virtually by definition a multiple-use and multiple-user resource.

Water belongs to users. If water is not used today it may be lost tomorrow, so much waste is encouraged. If water unused is a claim forgone, irrigators cannot recover the value of repairs to their canals, often built with assistance from the federal government. The legal definition of the right to water fosters inefficiency.

Because the apportionment of water in the arid West was tantamount to the allocation of economic survival, water law is rigid and resistant to change. This fact contains opportunities as well as problems. Water rights claims are layered by age: the oldest claim on a stream must be satisfied before the second-oldest can withdraw a drop. The Indian tribes of the Columbia basin are incontestably the earliest claimants to the river's water. But so long as the tribes lacked both capital to develop their own reservation lands for agriculture, and legal claims to the fish that lay at the center of their economy, their water rights meant little. Now the tribal lawsuits that have reaffirmed their treaty rights to fish harvest have also suggested a legal basis for assuring the productivity of the lands and waters upon which the tribes rely. The result is a looming claim, which the Indians have not sought to press directly, preferring the maneuvering room afforded by the uncertainty in the situation.

Thus uncertainty is a resource too. The Northwest Power Act and the fisheries rehabilitation that it authorized demonstrate the power of uncertain tribal claims. The possibility that dams might have to be removed to satisfy Indian treaty rights, or that other water users might find their claims foreclosed by the prior rights of the tribes, or that the tribes might take an equal place at the table with the state agencies in managing salmon—all these seemed far-fetched a generation ago. Now the last-mentioned event has already occurred, and that has motivated much change.

The perversities of water law result from the *proper* functioning of institutions, following the mandates assigned by legislatures or a constitution. A cure cannot be found in better implementation of policy or law, but in changing the law. That entails social and political action, which usually encounters resistance from those who benefit from the institutional arrangements as they are.

Changing water laws, which are rooted almost entirely in state statutes and regulations, will require crisis. The change, when it comes, will not be under the control of any single interest. Entrenched institutional bias often cannot be eroded, but must be overturned.

Organized Anarchy

The ideas discussed so far all assume that learning and action are instrumental, that they are aimed at achieving objectives that are known at the outset and do not change. But learning usually changes goals. In recognition of this fact, most American universities do not require students to select a major field until halfway through their undergraduate studies. In institutions, goals are frequently unclear, undermined by conflict, and remain unconnected to practical action. Actions often are undertaken as a way to search for goals.

Goals are highly fluid in the "organized anarchy" briefly described in Chapter 4. Action in an organized anarchy is a matter of timing: when a recognized problem has become politically urgent and when there is a plausible alternative to it, change, including dramatic change, is possible. The policy objectives pursued by government are sharply limited by the occasions that stir a sense of urgency—elections are times when political energy tends to be available—and the stock of alternatives and problems ready to hand when political opportunity knocks. The ability of government to perceive reality is dominated by the availability of the stimuli:

problems, alternative solutions, and political energy. These are shaped by institutions and their routines, including budget deadlines, elections, and the access provided to individuals—who is allowed to work on what issues. When institutional schedules or crises open the window of opportunity, the project that is ready is the one that gets acted upon.

The way institutions affect decision making in an organized anarchy is analogous to cognitive bias in individuals: the everyday understanding of what is important is sharply limited by the information actually perceived; it is often unrepresentative or wrong. Action in organized anarchy is opportunistic, and learning is episodic.

The inclusion of language on fish and wildlife in the Northwest Power Act demonstrates organized anarchy at work. Introduced in preliminary form in 1977, the act was known as "the power bill" because its focus was exclusively electric power. Then in 1980, at the request of an influential member of Congress from Michigan, a new section defining the fish and wildlife rehabilitation effort was suddenly inserted. That language stayed in the act, providing a vehicle for solving a problem (Indian treaty rights) that was essentially separate from power questions. The utilities, which wanted no additional claimants on electric power revenues, fought to have the section removed or watered down, but the link between the crisis in electric power and the crisis in salmon had been forged.

Even without the fluid uncertainty of organized anarchy, most organizations have deviant subcultures and professional cliques. Each kind of informal social structure remembers and understands reality differently from the mainstream. Each can become a source of learning, although in most cases that potential requires crisis before it can be realized.

The implication of organized anarchy is that institutional rhythms affect and may shape the way policy is made. Because timing is critical, actors must be ready before the battle starts: alternatives that could plausibly solve or address important problems need to be analyzed, so that their strengths and weaknesses can be quickly assessed; political leaders and their key assistants need to be sounded out about their receptivity to advancing a particular solution; and the window of opportunity to make policy, once open, must be entered quickly.

What Can Real Institutions Learn?

By surveying models of learning in individuals and in different kinds of institutions, we can develop expectations about what kinds of learning are realistic in organized human endeavor. As in the empirical survey of Chapter 5, the news is sobering but not all bad. Table 6–1 summarizes the points made in this chapter.

These social scientific models of learning encourage pessimism. The rational model, which defines learning, sets a standard that few

TABLE 6–1. Learning and expectations from nine models.

Learning Mode	Examples	Expectations
Rational		
Updating probabilities and values of alternative choices in light of experience.	Discovery that energy efficiency is the cost-effective "supply" of power.	Managers can be held to account. Reforms should overcome barriers to rationality.
Principal-agent		
Changes in behavior resulting from changes in ability of principals to hold agents accountable.	Substitution of hatcheries for dwindling natural habitat.	Individual rationality can lead to collectively undesirable outcomes. The harder it is to monitor the activity of the agent, the more likely it is there will be conflict between agent and principal. Faced with unfamiliar circumstances, mitigations should be devised by structuring the incentives faced by agents and principals so as to lower costs of information gathering.
Cybernetic		
Locating boundaries where critical variables go out of range, and managing so as to avoid those boundaries.	Projected power deficits signaled need to build new generating plants. Flow modifications benefit migrating salmon.	Implications of particular actions for larger goals are not taken into account. Learning may be super-stitious, particularly when objectives depend on outcomes.

TABLE 6−1 *(continued)*

Learning Mode	Examples	Expectations
		Skeptical experimentation (such as adaptive management) is both valuable and controversial when superstitious learning is most likely.
Coalition-bound incrementalism Finding solutions acceptable to all members of dominant coalition, usually by developing proposals similar to existing routines.	Evolution of water releases and dam operations to benefit migrating fish.	Incremental learning tends to be cybernetic and single-loop, with the limited capability to learn expected from both. Likely to be prevalent in large ecosystems, where cooperation, if possible at all, will take the form of a coalition.
Single-loop Solving problems within the conceptual framework of the organization.	Sockeye spawning channels (successful). Adjustment to sharing management responsibility with Indian tribes initially (unsuccessful).	Some learning can be made routine without threatening managerial control. Problems that do not fit the operating theory of the organization cannot be solved.
Double-loop Solving problems by reexamining premises and goals of organized cooperation.	Joint tribal-state fisheries management.	Some learning requires conflict or threatens loss of managerial control. When successful, real learning occurs—institutional theories of reality are improved, and the cybernetic disconnection of goals can be revised into a more accurate reflection of overall priorities.
Heuristic Skeptical appraisal of judgments inferred from everyday experience.	Effectiveness of fish hatcheries, judged on the basis of the first ones built.	In uncertain situations, guard against thinking something is not working. Type II errors

TABLE 6–1 (*continued*)

Learning Mode	Examples	Expectations
		are common and need the perspective of statistical analysis, gained through adaptive management.
Institutional constraints Institutionalized bias is rooted in mandates. Learning can occur through creative use of uncertainty to stimulate change.	Water rights are rigid but uncertain.	Uncertainty can be a resource. Large changes are likely not to be under the control of any interested party.
Organized anarchy Opportunistic action is the only tactical mode available. Change comes when windows of opportunity open.	Adoption of fish and wildlife language in Northwest Power Act.	Crises can be essential to change. Timing matters in such changes because responses to crisis are often not orderly. Those seeking change should prepare alternatives by tailoring them to the political agenda and getting likely leaders to be receptive to the ideas before crisis ripens.

other accounts approach, although the turbulent self-examination of double-loop learning would pass muster if it were used more often. Limitations on attention and cognitive biases skew the ability of individuals and organizations to learn from experience. Knowing there is a problem, however, is also the first step in finding ways to deal with it.

Comparison of the theories in Table 6–1 suggests additional lessons. Several have to do with incentives and learning:

- All the models describe institutional change, including learning, as the result of individuals' actions.

- The theories describe how the behavior of individuals is shaped by incentives and biases of psychological and institutional origin. Incentives can be designed and changed.

- In particular, single-loop learning can be planned and programmed. As with other organizational objectives, management support and stimulation (principally by making use of learning) are essential. Limited but real learning can be institutionalized through bounded rationality.

- Some changes do not involve learning: cybernetic attention to critical variables, superstition, heuristic inference, and institutional constraint, though all compatible with changes in organizational routines and institutional behavior, do not produce reliable understanding over time.

Other lessons have to do with conflict:

- A recurrent and important incentive is the threat to stability and control that arises from conflict.
- The rational learning model aims reforms at turning the decision-making process into closer approximations to the rational ideal. But several of the other theories point to the value of conflict for stimulating learning.
- There is an additional, surprising corollary: when conflict is stifled and the decision process is rigid, uncertainty can itself become a source of power.
- The more accurate and complete the operating theory of an institution, the more successful it is likely to be in responding to problems that could spark threatening conflict.
- It is unrealistic to expect that all conflicts can be anticipated and avoided. Double-loop learning and its potential for destabilization are unavoidable.
- Learning in the midst of conflict presents participants with two agendas: finding what is true and establishing who is in control. Coalitions are satisfied with incrementalism because small changes leave questions of power unasked. Organized anarchy relies on forces outside a policy network to create the opportunity to change institutional direction. These substitute for double-loop learning but may only postpone conflict rather than resolving it.

If the effect of policy is to be that of rational learning, it is vital to recognize and to mitigate the limiting and skewing effects of inappropriate incentives and cognitively biased institutional arrangements. Scientific skepticism, statistical checks on perceived patterns, and openness to conflict are all needed. Thus one arrives at the combination of idealism about science and pragmatism about politics that I have called social learning.

Patient Learning

One last lesson remains. The occurrence of learning and change does not guarantee that the understanding and institutional arrangements that emerge will serve better in managing an ecosystem. The important question is whether organizational learning leads to intelligent organizations, because it may not.

First, the past is not a good model of the future: studies of individuals' biases indicate that the experimental designs of ordinary life do not facilitate accurate causal inference. Second, organizations do not face a rich experience; bounded rationality succeeds in turning complex reality into standardized procedures and complex people into workers following rules. People participate effectively in teams when they work under conditions much simpler than those that prevail in everyday life outside the workplace door. The independent-minded person who shows up for work when he or she wants to is seldom a successful organization member. As a result, however, organizational experience tends toward simplified and biased samples, in which those who march to their own drummers are underrepresented. Third, *successful* learning extinguishes experimentation: this is the competency trap—"if it ain't broke, don't fix it."

An essential response is to be patient: let the base of experience get larger; take comfort in the realization that slow learners avoid competency traps better; make use of experimentation and errors; make large changes rather than small ones; but change one thing at a time so that the chance of seeing cause and effect clearly is improved. These ideas inform the hope for social learning.

Sustainable development must be both true in the sense that it should preserve the productive capability of the environment, and just in the sense that people should consent to the governance required to maintain sustainability. Thus learning, the search for truth, encounters politics, the search for justice. That is the topic of the final chapter. First, however, I must take up two less lofty issues. Social learning occurs through individual careers. Would anyone want to pursue social learning about the environment for a living? If one did, what would one do?

Chapter 7

Seaworthiness:
Civic Science

> It is not the business of political philosophy and
> science to determine what the state in general should
> or must be. What they may do is to aid in the
> creation of methods such that experimentation may
> go on less blindly, less at the mercy of accident, more
> intelligently, so that men may learn from their errors
> and profit by their successes.
>
> —John Dewey, *The Public and Its Problems*

Managing large ecosystems should rely not merely on science, but
on *civic* science; it should be irreducibly public in the way responsibilities are exercised, intrinsically technical, and open to learning
from errors and profiting from successes. Policies to learn must
persist for times of biological significance, and they must affect
human action on the scale of ecosystems. These are demanding
conditions, difficult to assure even if governments had unlimited
resources and needed only vision and will. In fact, of course, the
degree of control and the magnitude of resources available are always scarce. The strategy I urge—to be idealistic about science and
pragmatic about politics—is meant to respond to these real conditions. In reality a strategy cannot limit its attention to institutional
design. A strategy must identify that which is precarious and essential: In what ways must idealistic science be protected with stub-
bornness and guile? How can pragmatism maintain its civic virtues,
rather than drifting into opportunism?

The challenge of building and maintaining civic science and the
institutional relations necessary to do civic science is at bottom
individual. Civic science is a political activity; its spirit and value
depend upon the players, who make up, modify, implement, and

perhaps subvert the rules. The tenor of this chapter is therefore moral: how to recognize the dilemmas implicit in pursuing science in a political setting, and what values to protect in the compromises that we cannot evade.

Patience, determination, and optimism are values to nurture. Patience is essential because what needs doing will take biological time—maybe most of a career, maybe more. Sustainable development is likely to entail a long, slow change in the metabolism of the world economy. That is not to deny the urgency of pursuing sustainability. But the depth of the resistance and the breadth of the ignorance to be overcome mandate an indirect route. One moral choice, then, involves deciding whether the long-term goal is worth the short-term frustration and obstacles. Society as a whole need not face this choice—the long-term goal is an inevitable necessity regardless of short-term reversals and defections. For the individual, however, with only one career to pursue at a time, the long run may not be the main chance.

Determination will be needed in abundant measure. Learning is a precarious value—one that today's institutions find difficult to acknowledge in practice, however much they may support it in principle. Indeed, modifying and building institutions so that they can recognize and support long-term learning pose significant challenges: loss of funding for crucial information gathering, lack of job security, organizational indifference, shifts of fashion.

Those are the risks involved in most creative careers, however. The optimism springs from knowing that the steady accretion of perspective about large ecosystems can be won, and that such knowledge is likely to be used to reshape the human relationship with the natural world. These tasks lie at the outer limits of our experience, but there is no reason to think that they lie beyond our capabilities.

Combining large-scale science with political uncertainties nevertheless remains hazardous. We are used to thinking of power as a choice between doing good and doing evil; we are used to thinking of science as the search for truths that are inevitably beneficial. The conditions of large ecosystems force us to reconsider these simple notions, and to search for morally defensible ways of making compromises with reality.

We can address this mix of aims by borrowing a framework for appraisal developed by Garry Brewer, a scholar of policy sciences who has focused on the unstable chemistry of rational purpose and human means. According to Brewer, policies should be judged on four dimensions: (1) conceptual soundness, (2) equity, (3) technical efficiency, and (4) practicability. To these dimensions I add (5) vulnerabilities, the limitations and cautions necessary because learning in large ecosystems is not an established policy but an idea whose feasibility is being assessed.

Conceptual Soundness

Civic science is experimental science but reformist policy. That combination is radically vulnerable. Large-scale learning about nature cannot succeed unless it persists over space and time long enough to span many human conflicts; if those conflicts disrupt the learning process, the knowledge to be gained will be distorted or lost. Conversely, learning is never entirely comforting to power, because learning identifies error. Under what circumstances can these tensions be reconciled?

The Spectrum from Truth to Power

If the tension between science and politics were not deeply rooted, science and politics would not be such divergent human occupations. Understanding the divergence and the tension it causes is basic to seeing why learning must be at once idealistic and pragmatic.

Science and politics serve different purposes. Politics aims at the responsible use of power; in a democracy, "responsible" means accountable, eventually to voters. Science aims at finding truths—results that withstand the scrutiny of one's fellow scientists. These are ideals and goals, important in defining a moral order. In an industrial society there are two additional roles that combine knowledge, power, and morality, here labeled "administrator" and "professional/analyst," the latter including those, such as doctors or engineers, whose expertise is based upon science but whose practice is oriented toward the responsible exercise of professional judgment.

The idea of a spectrum of occupational roles and social functions (Fig. 7–1) was originally proposed by Don K. Price to identify the distinctive contributions made by different groups to a technological society.[1] Here, I want to stress the implication for individuals of

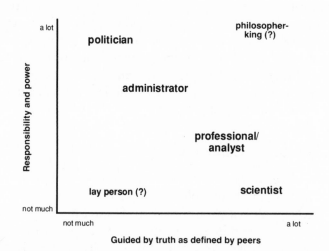

FIGURE 7−1. The spectrum from truth to power (after Price 1965).

the fact that the roles are separate. *A single person cannot play several different roles at once.* Any single person's probability of achieving success depends upon competition—what is called "fitness" in biology—and only very rarely does any single person achieve high levels of success in more than one arena.

This does not mean that one must live out one's career in a single niche. Certainly there appear to be successful persons in bordering regions where different roles overlap: some scientists advise Congress regularly; political leaders rely upon analysts, and most distinguish among them on the basis of merit as well as on the basis of political loyalty. But most people are trained for only one of these roles, and there *are* costs and risks involved in moving from one arena to another. If costs of movement did not exist, we would observe persons who combined excellence in science with success in politics, administration, and professional practice—persons I have labeled "philosopher-kings," after the wise and powerful governors of Plato's *Republic*. A few paragons of this type are reputed to exist, although closer inspection reveals that most of them acquired their reputations in some of these areas after they had already gained prestige in another. That raises the question of whether their later success was their own, or instead the reflection of a good staff. More important, neither our academic nor governmental systems rely upon or plan for these paragons. In a world of bounded rationality, one should allow for genius but not count upon it. Indeed, in a world of virtually universal specialization, most of us are not experts at all in most spheres of life; we are laity in the lower left corner of Figure 7–1, with power that reaches not far beyond ourselves, dependent on others' knowledge of what is true in most of what we consume or use.

Faustian Bargains

The spectrum from truth to power places a crucial constraint on civic science: in learning to manage large ecosystems we cannot rely on philosopher-kings. So there must be a partnership between the science of ecosystems and the political tasks of governing. As in any partnership, the relationship between principal and agent is inevitably problematic at some points.

Ecosystem-scale science requires political support to be done. Ecosystem management is now the responsibility of numerous public agencies and corporate property owners. Although they often act in a fragmented fashion that hardly merits the label "management," any management worthy of the term must be carried out by a substantial organization, probably a government agency or one subject to extensive public scrutiny. Learning in such a setting cannot take place without active political support; there are too many ways for things to go wrong without it. This is why political pragmatism is centrally important to those who wish to work scientifically at the ecosystem scale.

Why should politicians support adaptive learning? Learning is a responsible use of power, particularly so long as public enthusiasm for environmental protection persists. Political acquiescence is not difficult to win, but acquiescence is not the same thing as support. Support can come from a variety of sources: from an individual politician's personal commitment to the idea; from the expectation that learning will provide direct or indirect political benefits at some time in the future; from a perceived duty to implement a policy, perhaps combined with professional pride in seeing it done in an imaginative way.

Support can be extinguished easily, however. Once scientific findings point to the need for major reallocations of benefits or authorities, the politician is faced with the problem of achieving these changes. Few politicians are powerful enough to contemplate reform for its own sake—and reforms suggested by research scientists command attention on few grounds other than merit. This is not to say that politicians are cynical about intellectual merit. It is to say that their task is one in which intellectual merit is not compelling until the political ground has been prepared. Before that, findings may simply be irritants.

For this reason, and because civic science must have political support, *civic science should have an indirect bearing on policy* as much as can be foreseen.[2] That is, the lessons anticipated from adaptive management should *not* require double-loop learning. From a political standpoint civic science should be often a source of solutions, sometimes a source of single-loop learning, and never a challenge to legitimacy. Prudent politicians will not proceed in any other way. Even then, there is risk because their control over their agents is imperfect. Scientists who insist on different arrangements will not be supported at the scale and for the lengths of time necessary to learn at the ecosystem level. A comparative study of international relations and the use of scientific information by Edward Miles, a student of international relations, confirms this pattern.

To see how the dynamic works concretely, consider an experimental program to compare the relative effectiveness of hatcheries. Unless all hatcheries are making positive contributions to the stocks they are producing, research of this kind will increase pressure from advocates of wild fish to close down some operating facilities. A proposal to close hatcheries will predictably encounter fierce resistance from hatchery staffs, their allies within government, and fishing groups who believe they benefit from the targeted facilities.[3] It may be that senior agency officials believe that some hatcheries are ineffective or even harmful; it may be that the money saved could be put to better use within the agency. But so long as there is a risk that a high political price will be paid, and a risk that the hatchery will not be closed after the dust has settled, trying to close a hatchery can easily look imprudent. Under those conditions, "help" from science is unwanted.

Ecosystems are interconnected in ways that are not obvious to politicians, however. For example, detailed study of survival as fish migrate downriver past the dams produces information that measures the biological effectiveness of hatcheries in the Columbia, since each hatchery's production of juvenile fish is released somewhere and the identity of the fish needs to be known in order to study survival. It is a short step from there to combining information on the costs of operation of hatcheries, which is accessible because the funding is public, to produce cost-effectiveness ratings. Such information can be used in ways that are politically safe—such as waiting for a budget downturn to make selective cuts—if scientists and politicians are diplomatic within their own communities.

The experimental study of fish mortality is indirect in its effect on hatcheries, but it may threaten other groups battling over flows of river water. They will attempt to tilt the results in their favor. In general, the more relevant an expected result is to an important political choice, the more turbulence should be expected.

Practitioners of civic science must therefore take account of the political implications of expected results, but experiments can also surprise. A prudent scientist looks to the long run in dealing with surprises that are tied to policy. First, a scientist's political capital is his or her reputation; in that respect, both good and bad news must be reported in timely fashion, or else the scientist's ability to make further credible contributions will be jeopardized. Second, the time scale of politics is shorter than that of scholarly publication; often a year or more passes between submission of an article and its appearance in print. Timely warnings can take most of the sting out of public disclosure of surprise. The scientist needs to think carefully about whom to warn, however, since current political supporters may not be the only allies he or she will need over time. Third, the more salient the policy implications of a finding are, the more important it is to publish the work in a scientific journal. Publication serves as public confirmation of the soundness of the methods used, a valuable asset when there is controversy.

Maintaining an indirect linkage between science and policy aims at lowering the risk and cost of civic science to political supporters while decreasing the temptation of governmental authorities to interfere with the conduct of science. That approach also emphasizes long-term, system-scale learning, where our understanding of ecosystems is weakest. For both these reasons, the prudent approach husbands the scarce political support without which learning fails.

Experimental Science and Reformist Politics

The theoretical basis of civic science, in sum, has two parts, one technical and one social. Its scientific component is described in Chapter 3: controlled experimental manipulations of large ecosystems and ecosystem components, in which internal validity and power of test have been assessed, and in which external validity is checked against the existing understanding of this and other ecosystems. Although the conceptual requirements are imposing, the demands of time and resources are reasonable by comparison with the costs of carrying out major rehabilitative or regulatory policies at the ecosystem scale.

Civic science is reformist in two ways. First, it relies upon those charged with the operation of large technological systems to perform most of the experimental protocol and data collection. As discussed in Chapter 3, this implies winning and maintaining the confidence of operating organizations. In that sense, experimenters pursuing civic science cannot challenge the basic legitimacy of these operating organizations. Yet civic science cannot move toward sustainability without laying the basis for substantial changes in the level, scope, and organization of what operating agencies do in large ecosystems. That tension, between the legitimacy of today's ecosystem managers and the changes implicit in sustainable use, is an inherent diplomatic challenge of civic science. Working with those in operating organizations does not require sharing their views or goals; it requires being honest and thoughtful and inventive.

Second, in a political sense civic science should inform fundamental change while working from within. In the United States, working within the existing political culture involves acknowledging the substantial influence of nongovernmental organizations backed by environmentalists and by development interests such as business firms and labor unions. It involves recognizing the constraints and opportunities of divided powers within a federal-state governmental structure, which is further fragmented by the functional divisions and bureaucratic autonomy of major public agencies. This structure has been supportive of civic science, and a reformist tilt is a sensible extrapolation from recent history. It is less clear what to do elsewhere: with governmental secrecy of the kind found in most political systems; with political repression; with corporatist links between government and the principal institutions of the economy. Each of these can thwart the external enabling forces described in Chapter 4. But environmental progress is not logically dependent upon liberal democratic politics, and there is no compelling reason to think that sustainable development can be pursued only within a single political framework. On these matters, more experience is needed.

Equity

The criterion of equity requires an appraisal of two categories of people whose lives may be directly affected by civic science; those trying to do civic science, the experimenters and their political supporters; and those who are losers under today's resource utilization schemes. Here I also examine the question of when those doing civic science should take part in dispute resolution, an apparently promising way to shift the balance among winners and losers.

Professionalism

Learning at the ecosystem scale is a precarious value, something that no one opposes on principle but that tends to be eroded when it goes against the habits and biases of operating organizations.[4] Learning can threaten the operators of large technological systems, but those operators are necessary to carry out the experiments by which learning takes place. Even if the operating staffs believe learning is a good idea, they will be tempted to interfere. But they will interfere anyway, so what is to be learned and what is being learned must be shared with them.

Learning, like other rationalist policies, relies on professionals. Recall the epistemic community in the Mediterranean: their shared belief in science as the standard against which policy should be judged led them to coordinate their political activities across national boundaries, providing the glue that held together the Mediterranean Action Plan.

The importance of that epistemic community suggests that adaptive management can be nurtured by policy-oriented learning, described in Chapter 4. Its driving force is the advocacy coalition, populated by professional analysts and managers. They work in action-oriented organizational situations, using both political skills and technical expertise to advance political debate and to season ideas so that they become policy.

As is the case with all large-scale activities in technological societies, it is the professionals who reap the initial benefits. And the ever-present temptation to substitute self-interest for dispassionate professionalism creates a morally ambivalent situation. More demoralizing still, because it must be countered in everyday routine, is the reality that the office work and bureaucratic politics of civic science can seem far removed from sustainable use of the biosphere.

Political pragmatism is essential because there may not prove to be a way to work on the problems of large-scale learning except within the existing order. If one believes in the long-run importance of learning on a large scale, the means cannot be detached from the ends.

There are corresponding responsibilities: to energize and cultivate an epistemic community that can foster learning at the ecosystem scale across organizations and across lines of policy debate. The epistemic community constitutes the political support of those who carry out experiments in large ecosystems. Community members need to foster and protect learning if it is to survive. Members of the epistemic community who are staff members will find that this approach sometimes puts them in a difficult position: the person for whom they work may find an experimental result disadvantageous and wish to attack the experimenter, but the staff person must sustain a professional obligation to keep the pursuit of reliable knowledge on track. This is a principal-agent problem seen from the agent's point of view.

Can this be done? An epistemic community has emerged in the Columbia basin. Another consists of the environmental analysts hired by the federal government and interest groups when the courts made environmental impact statements the cornerstone of environmental law.[5] These impact statements improved federal decision making primarily because of the presence of outsiders watching how analysts and their agencies did their jobs. The outsiders had different competencies and powers: other federal agencies are either authorized or required to comment on draft statements, but they cannot sue; citizens' groups can comment and sue but generally lack the resources to do more than go after what they consider the worst cases; and federal judges can take what lawyers call a "hard look" at statements brought before them, but they react only to the cases on which suits are filed. The diversity of institutions thus linked into an epistemic community provides more strength than any of them could alone.

It is hard to forecast who the outsiders scrutinizing adaptive management will be. If the experiments appear to be irrelevant to decision making, the natural audience will be scientists interested in understanding large ecosystems—a professional peer group that is likely to be small and uninterested in pressing vigorously for policy changes. If the experiments seem to be directly relevant to decision making, the experimenters will face grudging cooperation and occasional sabotage from operating staffs. In either case, outside support that can foster good science is likely to be inconsistent.

Yet sporadic attention, spurred by the surprises that emerge, is quite likely, and can make a substantial contribution. Both the experimenters and their political supporters have an incentive to have good science prevail when a surprise comes along. To do otherwise would undermine the integrity of the experimenters and sacrifice the political high ground of their supporters. The controversy and attention aroused by surprises can have positive effects on both experimentation and policy.

In sum, learning is likely to remain a precarious value—successful in some places, perhaps in many, but not everywhere. Learning must first be woven into the fabric of the institutions managing the ecosystem—and then it must become an inconspicuous part of the pattern, so that reliable knowledge can be gathered over periods when leaders will come and go, budgets may increase and decrease, and public concern for the environment or any particular ecosystem may wax and wane.

Today's Victims

Civic science cannot involve learning over long times while simultaneously challenging governments with jurisdiction over large ecosystems. Therefore, civic science cannot by itself succor losers in the existing distribution of power and wealth. In contrast to the case of Albert Einstein, who wrote to Franklin Roosevelt to urge the development of the atomic bomb, the participant in civic science should not imagine herself on the front lines fighting for social justice through ecological learning. The knowledge that comes from adaptive management may come too late to benefit those now unjustly oppressed in the use of natural resources. Of the compromises that seem necessary to pursue civic science at all, acquiescence in established governments and their shortcomings is one of the most troubling.

Some unsustainable uses of natural resources are part of a pattern of across-the-board exploitation, in which governments are complicit or participate actively in the perpetration of human and environmental evil. In these cases, civic science cannot begin until political change has created a stable civil order. Opposition and the fostering of political crisis may be the only way to salvage anything of the biological and cultural heritages of such communities. This is political action—which may yield lessons but not systematic science. When governments are arguably reformable, individuals face uneasy political choices. Scientists have sometimes rationalized working with authoritarian governments in order to save tropical species, on the theory that there is no intrinsic connection between policies toward people and policies toward nonhuman species. In the context of sustainable development, this is an implausible idea.

Civic Science and Dispute Resolution

Adaptive management and conflict are complementary; each can catch errors and misunderstandings that the other cannot. Civic science entails both. But how should the differing needs of experimentation and dispute intervention be met?

Working at learning as if it could be divorced entirely from conflicting views produces a conflict of interest but not necessarily a conflict. As analyzed in Chapter 4, the aim of dispute resolution is to build a relationship in which the differing values of the parties persist, but in a way that promotes or preserves large-scale learning. There may not be an actual conflict, unless the disputing relationship sacrifices the interests of one or more parties to keep in place a long-term experiment. Such circumstances are clearly imaginable; for example, the indirect connection between studies of fish survival and hatchery effectiveness, discussed above, could be disadvantageous to agencies administering hatcheries of low or negative effectiveness. But in that situation it is hard to imagine that the disputing parties would not be aware of the implications of the accumulating knowledge. As a practical matter, the reverse effect is more likely: because of their apparently conflicting interest, the disputing parties may distrust those trying to establish a framework for civic science and forestall learning altogether.

The equity appraisal highlights the frailty of civic science rather than any potential it may have to do harm.

Technical Efficiency

Social learning works best when it produces good science: controlled experimentation, replication of surprising and important results, skepticism about cognitive biases in the conclusions reached—the elements of technically sound, efficient learning. Exercising political pragmatism in defending these ideals is therefore crucial to civic science.

Bias against Type II Errors

One way in which civic science makes a major break from conventional science is in its treatment of Type II errors—the rejection, as false, of propositions that turn out to be true. The search for sustainability requires scientific knowledge that promotes deep changes in the industrial order. That requirement entails a wider search for options. Conventional science has guarded against Type I errors— accepting as true a proposition that turns out to be false—but has paid little attention to Type II errors. Because safeguards against the two types of errors are partially incompatible, civic science includes as a guideline to experimental design the explicit assumption that *Type II errors are at least as deleterious as Type I errors.* Such a policy bias can take a number of forms. Guarding against Type II errors remains abstract until it is translated into policy recommendations; then it begins to look radical.

Burden of proof on the regulated. Rather than having government justify its regulation of those who use the environment, policy should shift toward requiring those who would intervene in environmental systems to bear the burden of proving that their interventions will *not* cause harms. The situation today is mixed. In some areas, such as the disposal of radioactive and chemical wastes, the burden of proof rests with the disposer; the government tends not to establish specific standards until some substantial quantity of waste is disposed of. In other areas, such as the protection of possibly endangered species, the burden rests squarely with those who seek protection.

Shifting the burden of proof all at once would be disruptive and would retard the formation of new businesses and the expansion of existing ones. The present hodgepodge is unsatisfactory too. Civic science points a direction but does not set a pace.

Permissive experimentation. Even though civic science values stable regulations, it also encourages experimentation. An experimental project should be given wide latitude, so that new hypotheses can be tested and explored. To be considered "experimental" a project should satisfy two conditions: it must be biologically reversible, and it must yield knowledge usable within the existing framework of understanding. The burden of persuasion in both cases rests with the would-be experimenter, to answer questions that are likely to arise about intervening in pristine ecosystems, which may need to be measured if they are to serve as control cases; major interventions in undisturbed areas are likely to fail the criterion of biological reversibility.

Data Bases and Models

The need to establish data bases and models to get a coherent picture of a large ecosystem was discussed in Chapter 3. Two other factors also affect their role in a technically efficient civic science.

Visualization. Understanding the behavior of large ecosystems is bound to be far from straightforward. Nothing is readily apparent about them: observations are scarce, theory is fragmentary, surprise is frequent. Accordingly, those seeking to manage large ecosystems would do well to make use of multiple ways of visualizing, sensing warnings from, and conceptualizing them.[6] Information as diverse as rates of fish harvest, sediment loads in a stream after rainstorms, satellite photographs, and marketing surveys can all supply useful and usable "sensations" with which to form a gestalt. These pictures of the complex reality are likely to be misleading and controversial. We need to use categories because our minds cannot keep track of all there is to know, even when all there is to know is too little to comprehend where the ecosystem is or where it is heading. But we should be wary of force-fitting information into categories. Instead, we should be skeptical of the categories we use, and forgiving of the errors that will inevitably follow.

In practice, this strategy entails using *both* field observations and summary methods sampling diverse sources and types of information both from the grassroots and from computer analyses. Excluding or dismissing information from any one source is risky, especially when those sources are likely to bring surprise. The danger is illustrated by an early computer-simulation model of electric power generation in the Columbia River: after plans were made on the basis of this model, an alert engineer discovered that the model assumed the river would flow uphill at night when power demand was low. It seems outlandish at first to put the correction of so gross a mistake under the bloodless heading "technical efficiency"; but the point is on target. Large ecosystems are easily misperceived; scientific discipline is needed as a safeguard.

Integrity and continuity. The most foreseeable risk to ecosystem management is that the overall picture of the system will be damaged by interruption of data collection as some measurements are discontinued and by loss of existing data.

Continuity must be a feature of policy. Monitoring should be justified and funded on grounds other than experimentation, because monitoring is too expensive to be defended solely on the basis of its contribution to learning, a precarious value. A continuous set of observations is too important to the ability to perceive the ecosystem to be left vulnerable. Monitoring is the environmental counterpart to financial accounting and reporting. It is logically part of the ordinary cost of doing business.[7] That fact is not appreciated today, in part because there are no generally accepted standards for environmental accounting. This is an opportunity for innovation.

Ideally, the monitoring protocol and data base design should incorporate ways of using monitoring data subsequently as control cases. But because it is impossible to know in advance what all the experimental treatments will be, it is hard to know how to design the control cases. Issues such as the spatial scale of collection, the species to be sampled, and the frequency of measurements are all potentially important. But not to make choices at this point is to ratify overcollecting, leaving monitoring vulnerable when budgets tighten.

A sensible strategy is to use the initial experiments to influence the shape of the data base and monitoring rules. Thus policies should encourage the development of as many of different *kinds* of experiments as possible at the outset—when enthusiasm and support are high and when there may be less competition for resources.

Expressed in a different way, the issue is this: At the outset, those seeking knowledge about an ecosystem lack a simple and clear set of principles about what data are indispensable to make the long-term dynamics of an ecosystem comprehensible. Defining such an ecological Rosetta Stone independently of specific ecological theories would help to bound the problem of how much and what kind of data are really needed on a regular basis.

Lacking such a conceptual foundation, we can apply one sensible rule: In times of stress, system knowledge should be protected. In principle, system knowledge should be accorded even higher priority than experiments that have been under way for years. When stress arises, the keepers of the data base will have potential conflicting interest with experimenters. Thinking these issues through before the stress hits would be helpful; stress will come soon enough, probably in the form of a lower budget.

Judgment of Threats

Campbell's injunction to analyze threats to experimental validity in a spirit of "evaluation, not rejection" bears repeating. The balance between the demands of ideal science and practical politics cannot be reduced to a formula or rule, however. Given the time and resources available, it is impossible to know enough to be sure that particular compromises will be tolerable, since in many cases such knowledge would depend on data that have yet to be collected.[8] The theory that might permit an educated guess is weak; that is why large-scale experiments have been undertaken in the first place. And values are involved; the judgment is not purely scientific, because the complications, like civic science itself, are social and political. Finally, the calculation of which threats to validity to respond to is subject to cognitive bias like other human judgments under uncertainty. Even when done well, judgment in deciding is only a start: courage in implementing and wisdom in changing course are needed too.

From a social perspective, the reality of these risks can be mitigated by having more experiments done in different large ecosystems with different kinds of jurisdictional complications.

A technical appraisal of civic science highlights the tension inherent in the attempt to combine rigorous science and practical politics. Civic science permits neither the comfortable illusion of detached objectivity of pure science, nor the intellectual sloppiness of imagining politics as the "art of the possible." The technical ideals of controlled experimentation and systematic skepticism must be defended by political adroitness. Practitioners of civic science will be evaluated both on how well they do in approximating the technical ideals, and on their adroitness in doing so.

Practicability

A pragmatic approach to civic science has four rules: (1) wait for crisis; (2) take advantage of disorder and slack; (3) be skeptical of the value of information; and (4) be patient.

The Uses of Crisis

Knowledge is important primarily to people who are painfully aware of their ignorance and confusion. Institutional interests, backed by cognitive bias and cybernetic decision making, pose formidable obstacles to the introduction of learning at the ecosystem scale. Most of the time those exercising power are aware that they could learn to do better but do not feel a strong urge to grasp that opportunity because they are already too busy.

Conversely, crisis can realign jurisdictions as unsuccessful organizations abandon the scene or are forced out. When a tree falls in a forest, new ground is exposed to sunlight. When a crisis arises, a new organization can be launched, as the Northwest Power Planning Council was to address the utility and fisheries problems of the Columbia basin. A new organization establishes its own institutional character: a style of approaching questions, a way of deploying its resources, a format for organizing its information, biases in what kinds of people to hire and how they should be trained. Each of these helps to alter the social ecology in which already-existing institutions operate. Over time, as the institutional character of the organizations sharing authority in the ecosystem adjusts, the definition of single-loop learning changes to accommodate new theories of how reality is put together, while forgetting or deemphasizing other principles.

Civic science affords an opportunity to expand single-loop learning to include a portion of double-loop learning: to raise the legitimacy of experimentation, and to lower the costs of being wrong. It is an opportunity that may not arise except when there is crisis.

Disorder and Slack

In a passage proposing the term *social learning*, the political scientist Hugh Heclo described the process as "a maze where the outlet is shifting and the walls are being constantly repatterned; where the subject is not one individual but a group bound together; where this group disagrees not only on how to get out but on whether getting out constitutes a satisfactory solution; where, finally, there is not one but a large number of such groups which keep getting in each other's way. Such is the setting for social learning." This description is remarkably similar to the description of organized anarchy in Chapters 4 and 6, except that organized anarchy focuses on *random* effects introduced by the chance overlap of several different streams of activity. It is not apparent how a model that is random can resemble a model of learning.

The connection lies in time frames. What Heclo means by social learning is something that occurs over decades, a process that includes much that could not be described step by step as learning but only the fashioning of dimly seen compromises under the press of circumstance. Organized anarchy was characterized earlier as "biochemical"—random connections in a loosely linked system. Large organisms are the aggregate of biochemical processes, however. What looks random at the cellular level looks full of purpose when the cells are organized as a flying bird. That is why something that feels anarchic to the bureaucrat on the scene can look, from the perspective of a century-long struggle over social welfare, like a well-organized pattern.

The pragmatic lesson is to seize the opportunity when it comes. Sustainable development does not have a definite timetable like an election. Finding means to move toward and settle into sustainable patterns of resource use will take a long time. There is no ideal time; there are mistakes to avoid, but no criterion defining the ingredients for sure success. If we had those we would be trying to do experiments of quite a different kind.

Value of Information

The problem with the economics of information is that one does not know the value of information precisely until after the costs of collection and analysis have been paid. If one is collecting repeatedly, as in long-term monitoring, the costs can be estimated with fair accuracy. For that reason, as well as because of the uncertainty about the questions one would want to ask of monitoring data, it makes sense to err on the side of collecting more than is absolutely needed. It also makes sense to keep collection costs low enough to survive a budget crisis. Unfortunately, as suggested above, these two principles do not necessarily have compatible solutions.

When experiments are involved, however, the bias should go the other way: to explore as many hypotheses as possible—which involves designing each experiment as inexpensively as possible. Given the scarcity of current knowledge, more can be learned from crude experiments probing *different aspects* of ecosystems than can be learned from refined measurements of theories that may turn out to be irrelevant even if they are correct. This guideline, then, is the diametric opposite of the one advanced for the data base. Data deficiencies need filling in even though doing so is costly. For example, one of the central problems of the attempt to conserve species is that so few species have been identified, particularly in tropical areas. This problem is one of data collection, and unfortunately has been a low priority among academic ecologists.

Experimental efficiency can generally be improved by tying the hypotheses to be tested to questions of *management* significance. Of course, doing that while pursuing the nearly opposite tack of keeping science only indirectly linked to political matters requires artistry. At the outset, however, there are so many questions of obvious scientific importance to be pursued—and by contrast such a restricted range of political controversy—that this is not a difficult hurdle.

Once hypotheses have been linked to management priorities, it is necessary to ask a question akin to the one asked by the Type II error test: how little data would be required to test this hypothesis at a level of significance *relevant to managers*? The last words are crucial. Management decisions often involve judgments that something is more likely than not to occur. That kind of judgment asks for information that reduces the probability of error to under 50 percent. This approach is drastically different from the conventional scientific one, which seeks to avoid Type I errors at a statistical level of assurance of more than 95 percent.

The Yakima hatchery now being designed illustrates the magnitudes involved. Its capital cost, estimated at $32 million in 1990, included $8 million for research. With the research costs included, the fish produced from this facility would be the second most expensive in the Columbia basin (and of course rising construction costs would make the price even higher). The price of knowledge from this facility looks high, especially by comparison with the Chelan County Public Utility District's East Bank hatchery not far away, whose cost per pound of fish produced is slightly under half that estimated for Yakima. Presented in these terms, the costs of research pose significant political dangers. Since the Yakima hatchery is the first full-fledged adaptive project in the Columbia basin program, its failure would threaten future efforts at adaptive management.

The research part of the budget does not specify the costs of education and science. The Yakima hatchery is more a laboratory than an experiment; the high costs come in an elaborate water-supply facility, intended to permit rearing of several different species of fish while avoiding spread of disease. East Bank has no laboratory features at all. One reason its fish are less expensive to produce is that there is no monitoring; consequently, it is impossible to tell if the hatchery is adding fish to the fishery, or if not, why not. Spending nothing for information in a situation of ignorance is not sensible policy.

But the Yakima design is also based upon statistical estimates that did not attempt to minimize cost. There was no apparent need to do so, since the hatchery was part of the Columbia basin program and there was only light review of cost figures—until the utilities introduced the comparison to East Bank. Then the cost became a political problem, but the design had already been done.

Information is seldom inexpensive when it must be collected data point by data point. Personal financial recordkeeping has been made relatively inexpensive because of extensive automation; even so, most consumers have no idea of the full cost of processing their bank checks. Trying to grasp the behavior of a large ecosystem is far more expensive. Hence the need for cost consciousness from the outset. This is a practical lesson of the Yakima hatchery experience.

The value of information needs to be apparent to policymakers, even at the cost of compromising the rule that science should be kept at arm's length from politically sensitive questions.

Patience

Biological systems show results slowly. Clever ways of observing living populations can be expected to emerge from ecosystem-level experimentation. But scientific technique remains quite undeveloped in this new field, and until those better methods are invented, political constraints are likely to influence the kind of learning gained. That is why an indirect link between science and politics is essential, along with patience. Patience flows from an appreciation that biological time scales require deliberate learning, and in turn stimulates efforts to keep resources focused long enough for that to happen. With patience comes awareness of the need not only to grasp opportunities, but also to wait for opportunities to ripen.

Vulnerabilities

There are several problems over which those doing civic science have little control. Taking precautions against such problems is valuable, like building a root cellar in tornado country.

Fiscal Crisis

Working on large ecosystems is expensive and is usually paid for by public funds. In times of extended fiscal stress, such as the U.S. public economy has experienced for more than a decade, research funds are cut quickly and restored slowly. In such conditions it is unclear how much large-scale research, and the social learning it stimulates, can be successfully justified as an essential part of ecosystem rehabilitation. Beyond that, there lies the question of whether ecosystem rehabilitation is a cost that governments are willing to bear, in competition with the other demands on a perpetually tight budget.

The likelihood of fiscal crisis at some point during long-term experimentation suggests that experiments should be designed with "soft-fail" features, so that no biological harm will result if they are interrupted and so that the data produced will have some value if collection is not completed.

Leadership Succession

In democratic governments, the political support essential for civic science is vulnerable to comparatively rapid turnover among politicians and political appointees. For this reason, practitioners of civic science need also to be teachers able to convey its importance to officials whose time is usually preempted by crisis and whose idea of learning is usually confined to the skills they acquire to survive on the job.

Change of Purpose

The time scale of ecosystem experiments may be greater than the life span of the social objectives that gave rise to them. Such objectives may change before data collection is complete. The environmental consciousness that has led us to think that sustainable development is an idea worth exploring dates only from 1970. Had we started taking data then, we might barely be reaching an understanding of how durable a contribution hatcheries make to the long-term welfare of the Columbia's salmon. Instead, it has taken more than 20 years to get to the starting line. Objectives that change as quickly as performance are fertile grounds for superstitious learning. That a program of ecosystem experimentation manages to win support from successive government administrators or to be interpreted as part of other, new doctrines may not mean the leaders of the program are succeeding at educating politicians and the public, but only that they have been fortunate so far.

Unbounded Conflict

In the context of pluralist politics, parties that were not included can ignite conflicts that cannot be contained, threatening the learning process. The fight over the preservation of declining salmon in the upper Columbia basin affords an example: a struggle that had been contained within the Northwest Power Planning Council planning process erupted in the larger context defined by the Endangered Species Act, an arena the council and its established disputants did not control. If the conflict cannot be returned to the council's framework, the vision of the Columbia as a system, and the hope that adaptive management could lead to reliable knowledge of that system, may be lost. The possibility that the Columbia basin program may be jeopardized less than three salmon lifetimes after it was established illustrates the fragility of ecosystem learning.

Risks and Visions

Large-scale learning has not been done before, in the systematic and intentional fashion advocated here. That may be because it has not been obviously needed. Or it may be because it is impossible. We cannot rule out the latter explanation without testing it: this is the largest experiment, one for which controls will be abundant but replication expensive.

If there is no guarantee of success, we are left to judge whether the prospective benefits to be gained, and the costs of not trying, make large-scale learning worthwhile. This constraint must be satisfied at the individual level as well as at the institutional level if adaptive policies are to be feasible. The choices involved may or may not prove tolerable for individuals. The risks do not appear to be greater than those of other life choices available to the well-educated. Challenges to institutions are substantial but perhaps not insuperable. Yet in the end, one runs risks not because they are acceptable but because the anticipated gains outweigh them.

Figure 7–2 reproduces the astronomer Woodruff Sullivan's *Earth at Night,* a photomontage assembled from dozens of weather-satellite images stored on magnetic tape near Boulder, Colorado. It does not show the real earth, but a cloudless image of the planet as it would look if it all faced away from the sun. With this artificial image, Professor Sullivan has created a valuable insight. His is a picture of decentralized diversity: from the forest-clearing fires of the tropical rain forests of Africa to the urban glow of California, from the natural-gas flares of the Arabian peninsula to the squid-luring lights of fishing boats in the Sea of Japan, humans have taken fire— the gift of Prometheus, who stole it from the gods—and put it to work in a thousand different ways.

Without the centralizing vision of science—without satellites to see the world—we cannot perceive the planet we share with other living things. But what we see when we look through the lens of science is a vast, decentralized, diverse reality that is unlikely to be reformed by central government because it is impossible to monitor in real time or to manage through central rule. That is the truth in the bumper sticker—"think globally, act locally."

The slogan suggests something else: if central *control* cannot work, the spread of ideas, brought to life by human intelligence in a decentralized fashion, *can* make a difference. There are two broad alternatives to central rule. One is a market; the other is a decentralized exercise of power—something that begins to look like democracy.

This image frames a definition of civic science as public policy. Civic science requires a democratic polity—one with electoral competition, open access to information, and free exercise of individual rights of expression. These ideals must be attained to a significant degree if the promise of democratic accountability is to be realized. Thomas Jefferson recommended watering the tree of liberty with the blood of tyrants. It must at least be possible to throw rascals out of office.

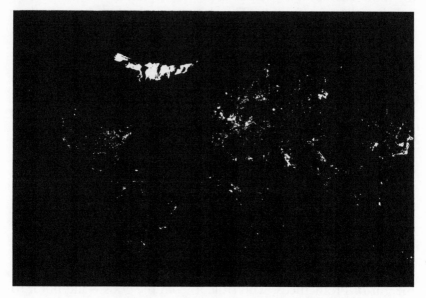

FIGURE 7−2. Earth at night. (Photo Woodruff T. Sullivan, III © 1985)

When science yields unambiguous negative implications for significant public values, action should be taken by governments. The form and content of the action will vary, but the imperative to act does not.

When science yields consensus on the importance of a problem, however ambiguous the existing knowledge may be, democratic governments should consider action, initiating or continuing the collection of relevant information and building their capacity to analyze that information over time. Scientific ambiguity is typically associated with political division, and the compromises shaped within the parameters of that ambiguity can be appropriate—but only if information-gathering is continued, so that evidences of error can be collected in a timely and potentially effective fashion.

Over time, such a stance provides as good assurance as can be obtained that the adaptations of industrial economies will adjust successfully to the changes in knowledge and in the condition of a planet on which human life has become a significant, though not controlling, force.

Chapter 8

Seeking Sustainability

... man, who at times dreamt of himself as a little
lower than the angels, has submitted to become the
servant and minister of nature. It still remains to be
seen whether the same actor can play both parts.

—Alfred North Whitehead, *Science and the Modern World*

The quintessential revolution is that of the spirit,
born of an intellectual conviction . . .

—Aung San Suu Kyi, *Freedom from Fear*

Searching for a Path

The preceding chapters have described learning as a means to
achieve sustainable uses of the natural environment. What have they
shown? First, that deliberate learning is possible, though surely
uncommon, in public policy. Second, that civic science—combining
a political strategy of bounded conflict with ecological learning
based upon experimentation—is feasible but fragile. Third, that
civic science promises the most rapid and least costly approach to
sustainability.

An emphasis on learning is an unconventional way to think about
public policy. Most policies hum with the tension between efficiency
and fairness—between maximizing some objective and sharing its
benefits equitably. Learning may not lead to the most efficient or
most equitable result, because fairness and efficiency are not enough
to define good policy when knowledge is sparse. Faced with ignorance, one can ignore it, hope to learn enough by trial and error to
muddle through, or try to overcome it. Although learning can
sometimes be deferred, the cost of information can never be evaded.

Once that cost is acknowledged to be necessary, civic science
becomes a "no regrets" approach: a strategy that yields benefits,

perhaps very large ones, at costs that are lower than those of muddling through. Even if civic science does not forestall all the crises of a world lurching toward sustainability, it will have enlarged the stock of knowledge. With determination and luck, that knowledge will be usable through and beyond dark ages, if they should come. Civic science is a strategy for policy and administration—a set of actions and values that has been shown to be feasible in some places, whose application elsewhere would move large ecosystems toward sustainability.

Yet strategy is not enough. Sustainable development is more than a problem of policy and administration. The changes brought by sustainability will redefine the place of our species in the natural order and inevitably alter the fabric of governance. Sustainable development accordingly presents fundamental political questions, questions about the character of human communities and what they mean by the common good. Sustainable development seems to promise affluence together with environmental quality. A society can be affluent, declared the anthropologist Marshall Sahlins, by having much or by wanting little. This is not a simple choice. It is also one we shall not avoid if we pursue sustainable development as well as talk about it.

The reach of markets and technology makes it necessary to think of a society of global scale, whose rhythms may be measured in biological time. Envisioning, building, and maintaining that sort of community will be an enormous challenge. My task here is to define that challenge more sharply, rather than to propose simple answers. The point is plainly stated: if continuing growth on the industrial model cannot be kept up forever, then we must ask sooner or later whether we seek to be affluent by having much or by wanting little.

This is a puzzling question. Its answer cannot be plotted with compass and gyroscope, because what is unclear is not only the path but also the destination. In addition to a navigator's map showing the way to sustainability, we need a settler's grasp of the topography of sustainability. The latter lies beyond the current state of knowledge because the sustainable economies we know from history all precede industrial technology and large-scale international trade. Sustainability includes three goals that overlap and partially conflict with one another. First, a sustainable economy must by definition respect environmental constraints and avoid irreversible damage to natural systems. In addition, a sustainable economy must be efficient enough to permit prosperity among a great majority of members of the society, and fair enough in its outcomes to maintain citizens' support for the governing order. Finding and retaining a balance among the aims of environmental soundness, efficiency, and fairness is difficult even in theory. My aim here is more modest: to identify three conditions that must exist on any path from the present to a sustainable future.[1]

That path must lead in two directions. Unless it brings a substantially better life for the poor who make up three-fourths of humankind, no path to development can be sustainable. This is the equity condition. Second, until we can forge a practicable link between our interests and those of our descendants, there can be no social basis for a claim to sustainability. This is the legacy condition. The third condition is continuity: the path to sustainability must begin from the present, inequitable and unsustainable world.

How can these conditions be met? The simple, daunting answer is that human societies will have to change the way they understand the relationship between economics and human life. Such a shift seems out of reach, and discussion of it utopian, until one realizes that it is already under way and, indeed, partly achieved.

Equity: Sustainable Parity

Intensive exploitation of resources takes place across historical time; the transformation of environments is the work of many hands, laboring for generations. The gulf that separates our postindustrial world from the rest of humanity has opened since Columbus sailed five centuries ago. One example of this process, from colonial America, gives a sense of the depth of the chasm and the possibilities of bridging it. The developing countries, like the Native Americans of New England, became poor as the first world became rich. This was not a simple process of extraction, but also a change of sensibility. One can be impoverished by having little or by wanting a lot. Poverty in the Third World partakes of both.

Humans inhabit a natural world from which they can obtain goods and services that they value.[2] But there are radically different ways for humans to use the natural world. Before Europeans arrived in the New World, Native Americans lived in hunting and gathering societies. Each year the people of New England migrated, following the abundance of nature through the changing seasons. Late in winter, when the food supply was lowest, the people edged toward famine, and the stress of malnutrition winnowed the human population.[3] The nomadic society lived lightly on its lands: human numbers were held down by the late-winter food supply, and the material consumption of each person was limited by the need to move each year. This was a stable, sustainable relationship—one the colonists called poor.

The improvements the colonists wanted came via markets in natural resources, which transformed the relationship between humans and environment. Markets link places. A large demand in Europe for American furs, timber, and other products of the land provided an opportunity to accumulate wealth that had not existed before contact with the Old World, when buyers and sellers did not differ greatly in their needs and access to the land and its products.[4] With contact the products of nature became commodities, things that could be sold for money. It was no longer sensible to take only what was needed, only as it was needed. Translatable into wealth and power, natural resources were exploited with increasing intensity until the ability of nature to support extraction was exhausted. Natural resources from America helped to fuel economic advances in both mother country and colony. The colonists prospered. And the life of the Native American changed forever, because its material foundations in the land were eroded and depleted.

Because places differ in their potentials for economic production, trade is often advantageous to both buyer and seller. When markets confer high values on some aspects of nature, such as the wood of tropical trees or the flesh of salmon, exploitation of timber or salmon may proceed rapidly, far outstripping the capability of natural populations to reproduce. Then those who depend upon the ecosystem for food, clothing, housing, and waste disposal lose their means of earning a living.

Today's developing countries are the inheritors of colonialism. Their lives are poor by postindustrial standards, and they cannot even return to a traditional way of life when colonialism has undermined the natural resource base of traditional economies. To a substantial degree the developing countries have no alternative but to develop— to find a way to live in a world remade by global markets.

The world refashioned by those markets is one in which wealth, income, and power are distributed unevenly. An important sign of inequality is the ability of some nations to export the costs of economic fluctuations to other, less powerful societies. Thus the prices of commodities harvested from developing countries rise and fall—sometimes with disastrous consequences to the producer nations, but usually with only minor ripples within consuming economies. The energy "crisis" of the 1970s was a rare turnabout. From the perspective of developing countries dependent upon sales of a few commodities, America, Europe, and Japan are perennially as menacing as OPEC was to us in the days of long lines at the gas station. Fluctuations in labor markets in North America and Europe wash waves of migrants out of and into developing countries. In America these waves have created a mythic shoreline lined with individual successes, with a cumulative benefit to the Canadian and American economies.

Protecting the biological heritage of weaker, poorer nations is likely to require either substantial redistribution of power and wealth among nations or the formation of dependency relations—some sort of colonialism—that can be stable over times of biological significance. In their absence, conflict over economic and social inequalities will continue to have environmental consequences, claiming both human and nonhuman victims. The requisite redistribution is large but likely to be affordable—as suggested by the recovery of Western Europe and Japan after the Second World War. Indeed, as the World Bank recently pointed out, environmental protection in developing countries is less costly than the current losses of income resulting from trade barriers and the large internal inefficiencies of their domestic economies. Furthermore, the overall objective is not *equality* of wealth or income, but a *sustainable parity* of economic condition. Today's foreign aid is an inadequately short step toward durable parity.

In sum, markets linking different, unequal places seem to foster unsustainable uses of nature. This is not a logically inevitable outcome, but the historical record suggests that it is so probable as to have been practically unavoidable—one more reason to be thankful for the end of colonialism. The reality that international markets often drive unsustainable exploitation of natural resources is an imposing barrier on the path toward equity.

Legacy: Unsustainable Growth

Those now living have inherited the world from past generations. That legacy includes economic growth over the past three centuries, propelled by technological advances and environmental transformations whose precedents can be found, if at all, in geological time. The advanced industrial nations have so far escaped the miserable, overcrowded poverty foreseen by Thomas Malthus at the end of the eighteenth century. Although populations have grown, the per capita consumption of goods and services has grown far faster, and we are wealthy in ways and to a degree undreamed of in earlier ages. Our wealth encompasses not only material goods but also qualitative changes in medical care, education, and political freedoms that have altered the terms in which we evaluate the quality of human life. The metamorphoses of industrialism are not uniformly good, and our response has been ambivalent in important respects. But the broad outlines of economic growth must still be called "progress."

Will growth continue? No one knows for sure. The traditional wisdom holds that because resources are finite, the appetites of a growing economy cannot expand indefinitely. The flaw in that reasoning is that we have found new resources (such as fertile soil in the New World and, more recently, oil beneath the North Sea) and new kinds of resources (such as silicon and uranium, both worthless until uses were found for them) fast enough to sustain economic growth for a long time. In the search for new resources, few geographic frontiers remain: the spread of European culture in its twentieth-century guise of industrialism has reached virtually all parts of the earth. Electromagnetic waves originated by humans carry signals throughout and beyond the solar system, and there are few places left where people live without the means to hear or see them. Economic growth has already caused irreparable damage such as the loss of valued species and the destruction of fertile lands. But the critical uncertainty lies in the fact that many *technological* landscapes remain to be explored.[5] A world still growing richer can further increase its options by investing in research and development. As with silicon and the amplifying importance of information, new technology may well draw upon resources that are plentiful and make changes that are extraordinary in speed, range, and consequence.

We do not know, then, when the Malthusian limits may be reached, because we have little insight into the mysteries of invention. Moreover, it is clear that in some poor countries, where the degradation and exploitation of natural resources are proceeding with little counterbalancing capital formation, Malthusian limits may have been exceeded already. Instead, we should ask what the implications are of continued growth on a global scale.[6] Here the answer is paradoxical: the worldwide spread of market economies, rightly hailed because they have been the engines of growth, cannot bring about sustainability without explicit intervention. Efficient markets promote the use of resources and ingenuity so as to increase material prosperity. Efficient markets also highlight the limited ability of governments to regulate behavior so that economic growth can improve the well-being of people. Well-being originates in economic growth, and growth requires efficient markets—but well-being also requires regulations that impair the efficiency of markets and may hamper growth. That is the paradox.

The tension between growth and well-being has a particularly stark implication for sustainable development: left to themselves, rational economic actors will not behave in ways that are sustainable. Instead, efficient markets tend to allocate resources to the current generation at the expense of later ones. Crucial to our understanding of this concept is a technical idea, the discount rate.

Investments are ways of putting time, energy, and other humanly controllable resources to work. Successful investments earn a positive return—a stream of benefits that, over time, more than repays the resources put in. Different investments yield returns in different ways; a racetrack wager may pay off in a few minutes, a forest only over a generation or longer. How should one compare these differences? The answer lies in the market. Investors make choices according to two factors: the rate of return of alternative investments and their riskiness. In general, the lower the risk, the lower the rate of return demanded by investors. The market-determined rates of return are affected by many factors that cause them to fluctuate, often by substantial degrees over a generation or longer. What matters for our purposes here is that market-determined rates of return exist and are almost always above zero.[7]

So long as the rate of return is above zero, the value of a resource in the future is lower than it is at present. This is a phenomenon known as discounting, and the pace at which the value of future gains is reduced is called the discount rate. If an investor can obtain $1,000 next year, she would accept less than $1,000 today because she can put that smaller sum into an investment that will increase to $1,000 in a year's time. The difference between $1,000 next year and an equivalent lesser amount today is the discount.[8]

The implications of this simple logic for natural resources are profound. Even at modest discount rates, values erode severely over periods of biological significance. At a discount rate of 2 percent, for example, $1 million in 100 years is worth $132,620 today, about one-seventh of its value; at 10 percent, $1 million in a century is worth only $30. Market rates of return have varied from about 0.5 percent to 9 percent, depending on riskiness, over the past 60 years; and contemporary practice requires rates of 10 percent or higher.[9]

The precise numbers are not the issue, since they will vary depending on the question to be answered. The phenomenon is fundamental: *if resources are traded in markets, the value of conserving them for ecologically significant lengths of time is set by markets, not by biology; usually, biological conservation turns out to be worth very little.*[10]

That is why it made sense for the colonists to ruin their land. Converting beavers into pelts and pelts into money permits the money to be invested. In that form, it will grow faster than beavers. This logic faced each trapper. By 1700 beavers had become scarce in New England. In the sweep of American history that depletion is little remembered: we had a continent to exploit, and by the time we began to reach its limits, the old landscapes of the colonies had regrown to look natural. Only a few people know that they are biologically much poorer. Why does this matter to sustainable development? The economic logic that transformed New England is still at work in the rain forests of Borneo and southeast Alaska. If resources are economic commodities, their value decreases over time; and overharvest, biologically speaking, can be economically rational.

To counter this logic requires at least one of four factors. The first is the indefinite continuation of technological progress, so that parts of nature that were valueless in the past become valuable resources, displacing those now being exploited. Thus the application of steam power to oceangoing ships made tall trees obsolete, since lofty masts were no longer needed to support sails; the use of coal to make steam saved the Douglas fir—and wiped out the investors who had gambled that their Douglas fir lands would be valuable. A continuing stream of new technologies cannot be relied upon. The economic role of technology does remind us, however, that sustainability is something different from preservation. Steady technological advance over the past three centuries has transformed the world both in a material sense and in the way we regard the natural world. There is no sign that this transformation has ceased.[11]

A second factor is property rights. If the Native Americans of New England or the colonists had had a guarantee that future harvests would be theirs by right and of stable value in the marketplace, they would have had economic reason to husband their resources. That is why sustainable management of the rain forest of northern Australia was plausible on economic grounds. This logic also operates implicitly in many preindustrial societies' notions of property: to take only what is necessary, so that there will be enough in the future. But it is precisely the instability induced by global markets and technological change that makes property rights an uncertain proposition over biologically significant lengths of time. Even more compelling is the fact that property rights are usually imperfect, so that those in a position to exploit a resource have little reason to conserve; they do not expect to benefit from gains in the distant future. The corrupt government officials of Southeast Asia who harvest tropical timber at rates far higher than can be sustained are being economically rational: they have reason to believe that their control will not last, so they must take their gains now if they are ever to have them. Property rights that induce lasting husbandry are an essential component of any policy for sustainability. But they are not self-evident or self-executing. We cannot look to an unfettered market to work this out on its own.

Two other factors seem more robust, at first, than fortuitous technological change or property rights: cultural values that limit the extent to which natural resources are treated as commodities, and governmental controls that limit the choices of economic actors. But in a global economy, societies that have or undertake such limitations become vulnerable, over time, to the political and military power of those that do not.[12] Explicitly arranging for insurance that compensates for the global risks of unsustainability could work, but that would entail an international agreement to limit consumption to the net income earned from the planet's assets—an idea that is reasonable but far from politically attainable today. Without such an agreement, worldwide exhaustion of resources will eventually lower the returns of *all* alternative investments, bringing down the discount rate and abating the economic pressure to overexploit. It may well be too late by then, however, if the pressure to meet immediate demand draws resources away from investment altogether. Burning dung for fuel instead using it as fertilizer in Ethiopia cost that nation's economy about $600 million per year in the early 1980s, 30 percent of the total value of agricultural production. But there was no other fuel. Burning the furniture to stay warm is unlikely to support well-being for long, but it is going on already.

The power of efficient markets is compelling both internally and externally. Its internal force can be seen not only in the verbal form in which I have put it above, but also in formal, mathematical demonstrations that positive discount rates in efficient markets transfer resources from future generations to those now living. The external force of economic efficiency has operated since 1492. No peoples have successfully resisted the expansion of the global market. Some, like Japan until 1868 and the socialist countries since 1917, have delayed the impact of global markets for substantial periods by the use of state powers. But their resistance has been unavailing, and the environmental costs of industrialization in the socialist nations seem, in any case, to have been greater per unit of gain than in the capitalist economies.

The shifting of economic benefits to the present that discounting causes can be justified only if economic growth continues. For economic growth in the present *can* enlarge the stock of knowledge and capital available to future generations. In a trickle-down economics that crosses generations, our descendants can benefit from our wealth today if we invest and learn so much that they will be wealthier than we. But only if they are wealthier—only if growth continues. Thus the discount rate fosters a kind of pyramid scheme. If growth halts, those who have invested in the expectation of growth will lose; it is just this fear of being caught standing when the music stops that suddenly shuts off demand and investment when economic slowdowns begin. But recessions, sometimes under the prodding of government and always fueled by new opportunities within the economy, have been temporary lulls in the long march of growth. So far.

Is the hope that growth will continue a plausible one? Analytically it is not: there is no guarantee that growth can or will continue. Moreover, the paradoxical intertwining of growth and efficiency increases the danger that growth will falter, since growth is fostered by efficient markets but sustainable economic activities are undervalued in efficient markets. Yet in a world of competing investors, it is compellingly rational to discount the future: the investor who refrains from discounting will do better than her unthinking competitors only if growth stops within the time period of their investments. That has been a bad bet for so long a time that the notion that one can be rich by wanting little is not even taken seriously by the institutions that steer societies and economies.

There is a kind of perverse Pascal's Wager at work here:[13] one can either believe in the necessity of a sustainable future or not. If one acts upon that belief, but others do not, the unsustainable economy may or may not crash. If it does not, one has sacrificed for nothing. If it does, one still cannot escape the consequences, to the extent that one's life is bound up with the unsustainable order. So one may as well not believe in the necessity of sustainability.

The barrier to the legacy condition is the socially constructed assumption that economic growth will continue. It is an assumption without analytical foundation, which induces behavior hastening the day when the assumption will be incorrect. Yet growth is an assumption from which individuals cannot simply back away. Rather, they must change belief and actions in enough people for enough time to change the trajectory of economic institutions. Like the redistribution needed to meet the equity condition, it is, perhaps unhappily, a political task to build a social commitment to buy insurance, so that sustainability can become possible for efficient, growing economies.

Continuity: Compacts and Colonialism

That sustainable development has a political dimension is obvious. The question is whether the political commitments necessary to achieve sustainability can be assembled and maintained.

There is no secure answer, in the sense in which the economic arguments above can be grounded in logic. The momentum of the unsustainable is so deeply rooted, in the industrial order and in the markets that now tie all lands together, that judgments based upon probability would be very likely to dismiss the goal of sustainability altogether. Yet to put one's faith in an as-yet-invisible consensus on the need for sacrifices to reach sustainability would be simply innocent or else fatuous. Is there another course?

If there is, it is likely to lie through the self-interested agency of rich nations and their support of international institutions, both public and voluntary. Only the already-rich have the wherewithal to peer into the future and to consider its uncertain prospects. The already-rich have the most to lose, not only in a material sense but also, more importantly, in a moral one: the prospect of simply leaving the poor to suffer is so politically punishing as to be unacceptable. Prosperous societies have multiple objectives, but they also share humanitarian aspirations. If the horror of human suffering as portrayed by the mass media does not slide further into solipsistic consumerism, much will be done.

From this point of view, sustainable development is a form of insurance, a way of using part of the wealth produced by economic growth in advanced industrial nations to search for and to carry out economic development that is able to stabilize situations teetering toward disaster. Adaptive management and the interest-group politics discussed earlier fit naturally into such a long-run response to global inequality.

From the standpoint of the poor nations, sustainability offers a social compact: to join in the civil struggle for a fair apportionment of well-being, even though the shares will not soon be equitable. This is not a good bargain. It may only be better than any other one available. The alternatives include wars of redistribution; spheres of influence centered on wealthy nations such as France or Japan, which may themselves remain committed to unsustainable practices; or continued dependency and decline, with limited possibilities for determined and lucky individuals to make the great ascent on their own, leaving their fellow sufferers behind. These alternatives have been explored and will continue to be exercised. Their common characteristic is that they all disregard the question of how sustainable relations between people and nature could be built. As difficult as that question appears, the alternative of ignoring it is tantamount to admitting that there has been no progress since "*après moi le déluge.*"

For the poor nations, sustainable development is a new kind of colonialism, possibly a kinder colonialism. As the industrial powers struggle to define an international order beyond the cold war, the developing countries too must find ways to move into a global configuration in which there are no longer first and second worlds.[14] One possibility is environmental imperialism—a pattern of economic relations that permits resource exploitation to continue in poor nations without destroying their indigenous economies. That may include a shift of manufacturing to developing countries, taking advantage of their low-wage labor pools and large number of potential consumers. That path also affords greater domestic peace in developed societies, with their growing aversion to both polluting facilities and immigrants. Such arrangements are likely to be perceived as the sort of "win-win" solution that can emerge from thoughtful negotiation.

Mutually advantageous or not, such arrangements are imperialist in a cultural sense, for they demand of poor nations that they accept the definition of the good life from the developed economies. Mass media have already brought such definitions to developing societies, persuading many but also inciting Islamic and other backlashes. For all developing societies, tradition is being transmuted by the radiating demands of world capitalism. Central to those accommodations is the acknowledgment that poor societies *are* poor, that what they lack is first and foremost material fulfillment. As the troubling history of Native Americans suggests, this acknowledgment may not be made easily. A critical element of democratic governance is that these accommodations be made by the affected communities, in ways they hold to be legitimate. This procedural ideal may prove to be as hard to realize as a self-determined acclimation to modernity.

Does sustainable development promise no better options than some form of colonialism? I have chosen not to return to the China my parents left behind. The choices of nations are less clear-cut. But the analysis presented here suggests that *if* modernity is the goal, then an adaptive search for sustainability should be the mode of its pursuit. Only by acknowledging our ignorance of the ways of nature and by engaging with ignorance systematically can we minimize the hardships of transition. For nations learning is a crucial means to find a path to sustainability.

The conditions of equity, legacy, and continuity frame a political choice but do not determine its outcome or dynamics. This is so for both practical and philosophical reasons. Pragmatically, we know far too little of the processes of social interaction to predict how or when change of this scale might occur, nor is it clear how sustainability, once begun in some places, can be spread elsewhere. Nevertheless, an understanding of the framing conditions can afford insight into the trajectories that are possible given the starting circumstances. Philosophically, the choices that lie ahead appear to be free in the sense that there is no compelling force that requires a particular result. One of the principal attributes of sustainability is its profound empirical obscurity.

Hopes

At the Ryoanji temple in Kyoto visitors can drink from *Tsukubai*, a seventeenth-century water basin famous throughout Japan; on it is inscribed, "I learn only to be contented." That is, one who learns to be contented is spiritually rich, while one who does not learn to be contented is spiritually poor even if materially wealthy. Societies can be rich when they have much or when they want little. If it becomes possible to be affluent in the sense of wanting little, then sustainability may be possible; otherwise, technological progress is our only hope.

Sustainable development is not a policy objective so much as it is a vision of appropriate human endeavor on the planet we inhabit.[15] Although in principle the pursuit of sustainability could be developed into a portfolio of policies, given the uncertainty of human action no such grand plan is practicable. Policies such as adaptive management must meet a test of feasibility: they must seem likely enough to work to be adopted, and they must be workable in practice if they are to make a difference. Sustainability need not meet a test of feasibility. Rather, it must meet a test of justice. Once economic growth falls behind population growth for enough of the world for long enough, we shall be visibly on the way to sustainability. The question is what can be done in advance to lessen suffering and to assure equity in the experience of that which cannot be avoided. Over the past century we have learned much from our social responses to economic insecurity in industrial societies. We have not eliminated poverty, although knowledge has made a difference in how we cope with the human condition through public means. The record is undeniably a mixed one. Whether we can do even that well in environmental policy remains to be seen.

Tests of justice are tests of the whole rather than of its parts. No single action can create or destroy sustainability. That is why lampooning the fussiness of the "politically correct" is an effective if cheap shot; the real test of environmental soundness is not whether any particular can is recycled. Those who reduce justice to its parts are readily accused of trivial zeal. This does not mean that it is irrelevant whether any cans are recycled. Rather, what counts is the whole, which may have integrity even though its parts are flawed. In that same sense, sustainability as an institutional value succeeds or fails at the systemic level, much as racial justice must be gauged not only by compliance with ethnic quotas but also by the decent behavior of individuals to one another.

Sustainable development also cannot be reduced to a recipe because we have neither a list of ingredients nor a kitchen. The complex task of renewing political institutions is not one to be undertaken lightly, as the sociologist Peter Berger recalled at the foot of a Mexican pyramid: "The great pyramid at Cholula provides a vision of a succession of theoretical schemes, each embodied in stone and superimposed upon successive generations of silent peasants. . . . The pyramid was not designed for aesthetic purposes. . . . The meaning of the pyramid was provided by its sacrificial platform, the theory behind which was cogent and implacable: If the gods were not regularly fed with human blood, the universe would fall apart." Values inscribed in social institutions exact human costs. Growth does so today, as certainly as did the sacrifices of the pyramid-builders. What we can be sure of, in a world of complexity and uncertainty, is that we do not know enough about how to reach sustainability to minimize the human costs of doing so. We are not yet even able to estimate whether the visions of sustainability that have been devised so far do better than unrestrained growth.

The second reason why sustainable development cannot be reduced to a formula is diversity. A strong case has been put forward here for a centralized vision of the planet. There is no corresponding case for central control. Adaptive management of large ecosystems is necessary as a way to search for sustainability. What no one knows is whether central governance is possible, or necessary, or sufficient. So sustainable development remains uncalibrated in a crucial dimension: the power needed to carry it out.

Patience

In the end, however much or little is done to move toward sustainable development, whether it will be enough—indeed, what meaning will attach to "enough"—is itself a grand experiment of planetary society. How much misery will it take to make a global norm of sustainability first visible, then credible, then feasible, then inevitable? We do not know. And we do not know if the lessons of environmental disaster can be learned in time to ward off still more suffering. However bleak that prospect, we in the rich nations must bear the certain knowledge that our societies are both historically responsible for many of the circumstances that imprison the poor and that we will on average fare much better than they.

Against this background it is possible to see that sustainable development is not a goal, not a condition likely to be attained on earth as we know it. Rather, it is more like freedom or justice, a direction in which we strive, along which we search for a life good enough to warrant our comforts. Freedom and justice are easily taken for granted, although many have died in their pursuit and defense. A more materialist goal such as sustainability is harder to imbue with romance and ideology. But the enormous changes our species has wrought leave us a difficult choice: either to accept our humanity in the company of the whole human race and the natural world we jointly share, or to concede that being human is too difficult for the richest, most advanced beings in history.

The catastrophic changes that befell the Native Americans of New England, like those that disembowel the Third World today, result from the ability of a colonizing power to shift the costs of development onto indigenous peoples and future generations. But we are all indigenous to the Earth; we have no other home; we cannot simply leave. It is this reason, finally, that impels us to learn at the deepest levels of both natural and social science.

Over the last 500 years humankind has become a planetary force. We are not as powerful as the weather or as patient as geology, but as species vanish from the Columbia River to East Africa, as the ozone thins and toxic chemicals seep through aquifers, it is increasingly clear that we intervene decisively in the natural world. The Earth as seen from space reminds us that our world can no longer be treated as a treasure to be plundered. The garden is an appropriate image for the Earth after Columbus—a place that is bounded and organic, designed yet open to seasons and elements, natural but cultivated, sustainable and humane.

We have not found our way to the garden yet. What we have are science and democracy, the compass and gyroscope that are the heritage of the Columbian odyssey. What we need still is a sense of the patience of Earth. A century after Columbus sailed, Francis Bacon called for humankind to understand nature in order to subdue nature. The message of sustainability is that we must acknowledge the pace and scale of nature's teaching. That will not be an easy message to grasp or to heed, but it is perhaps the central test of whether we shall turn out to be a species with the determination to cultivate a world in which we can live together.

Notes

1. Taking Measures

1. I return in the final chapter to the issues raised by this scenario. Economists rightly observe that depletion of resources is not logically a consequence of colonization; rather, depletion reflects poorly defined property rights. This is an important theme in the design of sustainable policies for the future, although it is hard to imagine that conquerors of new lands would have scrupled over the property rights of those they had just subdued.

2. See the exchange between Cronon (1990) and Worster (1990) for a discussion of the degree to which capitalism, in the sense described by Marx, is useful in analyzing the environmental transformation of North America. My argument is simpler: the establishment of commercial ties between the Old World and New changed the terms on which a sustainable equilibrium in the New World could be conceived, since trade would play an important role so long as there were economic factors favoring trade.

3. The very instability of commodity markets makes it rational to sustain a subsistence relationship with the natural environment. That this is rarely observed suggests that there is more to be understood than economic incentives.

2. Sustainability in the Columbia Basin

1. Worster (1985) was commenting on the Colorado River.

2. The figures in Figure 2–4 are for *firm* electricity use—power provided by utilities whenever customers turn their switches. Firm power generally commands a higher price than interruptible power, which is sold without a guarantee of its availability. Hydropower dams do not have an assured "fuel" supply—water from rain and snow. Hydropower systems are therefore planned around their firm generating capability, the amount of power expected to be available even under conditions of severe drought. In most years, when there is more water than that

minimum, hydro systems generate considerable quantities of interruptible power. Much of that power from the Columbia is sold to markets in California, as well as to Northwest industries.

3. In traditional utility practice, the cost of a new power plant is not counted as part of the "rate base," or invested capital, until it goes into service. Ratepayers are consequently shielded from the cost impact of utility managers' investment decisions until power is delivered; by that time, it is generally too late to affect the project. In the Northwest, the costs of the first three WPPSS nuclear plants were financed by Bonneville, but the cost impacts were delayed by the financing arrangements until 1979. Since then, even though only one of the plants is in service, the costs of all three are being recovered through Bonneville's rates.

4. The principal exception was the suburbanizing service territory surrounding Seattle, served by Puget Sound Power and Light, where the prolonged boom of Boeing's commercial airplane business and the rise of computer software, led by Microsoft, powered sustained expansion of jobs, population, and economic activity.

5. By this logic, customers who did *not* participate would benefit as well, since their rates would not rise as rapidly as they would otherwise. Nonparticipants are free riders: they run no risk and need not undergo the stress of change. The fact that inertia is rational in this case remains a challenge to policy implementation.

6. While the program was being implemented, concern rose sharply about deteriorating indoor air quality in houses that had been tightened for energy conservation. In order to offset the risk of worsening indoor air quality, the Hood River project included energy-efficient heat-exchanger ventilators in its package of recommended measures. These ventilators increased costs as they produced a more healthful indoor environment, but they reduced energy savings.

7. Hirst, Goetz, and Trumble (1987) suggest that the discrepancy can be largely explained by the choice of base year. In 1982–83 rates were undergoing a steep increase, and rising prices depressed electric power consumption. Nonetheless, statistical analysis of the Hood River data indicates that only 40 percent of the discrepancy can be explained in this way.

8. In 1989 state regulatory commissions in New England began to implement rules for conservation programs designed to overcome the problems of revenue losses. This innovation relies upon the growing credibility of conservation as a resource that can be acquired on terms similar to those of generation.

9. The first hatchery master plan, drawn up for the Yakima Production

Project in 1990, incorporated state-of-the-art designs and practices to protect the genetic pools of wild fish and to make rigorous learning possible.

10. The dotted lines in Figure 2–5 reflect a 95 percent confidence level—that is, there is a 95 percent probability that the true trend line will fall within the dotted lines. A lower confidence level would allow the dotted lines to fall closer together. The fact that the observed data are already close to the dotted lines, however, indicates that a more relaxed confidence level would not change the conclusion stated in the text. No statistical manipulation will make the noise in the data go away; only a longer series of observations can do that.

3. Compass: Adaptive Management

1. I am grateful to Jan Carpenter for assembling background information on system planning, which was used in developing Table 3–1.

2. I have slightly rearranged Campbell's threats to validity, which overlap to some extent, in order to highlight their relevance to policy. Two were omitted because they are not directly related to policy considerations: *experimental mortality*, an apparent effect that stems from different rates of loss from the experimental groups rather than from the effect claimed; and *selection-maturation interaction*, a selection bias that results in different levels of maturation in the treated and control groups. These potential threats *are* germane to experimental interpretation, of course, and should be taken into account from the design stage forward.

    The standard terminology of this subject arose in a medical context. In the setting of ecosystem management I have normally used *intervention* for the standard term *treatment*.

3. "One day when I was a junior medical student, a very important Boston surgeon visited the school and delivered a great treatise on a large number of patients who had undergone successful operations for vascular reconstruction. At the end of the lecture, a young student at the back of the room timidly asked, 'Do you have any controls?' Well, the great surgeon drew himself up to his full height, hit the desk, and said, 'Do you mean did I not operate on half of the patients?' The hall grew very quiet then. The voice at the back of the room very hesitantly replied, 'Yes, that's what I had in mind.' Then the visitor's fist really came down as he thundered, 'Of course not. That would have doomed half of them to their death.' God, it was quiet then, and one could scarcely hear the small voice ask, 'Which half?'" (Tufte 1974, p. 4—attributed to Dr. E. Peacock, Jr., chairman of surgery, University of Arizona college of medicine, in *Medical World News*, Sept. 1, 1974, p. 45.)

4. There is an entirely different approach, Bayesian statistical analysis. The

Bayesian framework puts the burden of judgment on the decision maker in a different form and at different stages of the analysis. For a non-technical introduction, see Berger and Berry (1988).

5. Often a random selection method can be more fair than one that heeds imperfect reason. But since politics is the alternative to random selection, the random approach needs a champion, and a situation that can be described as one in which the best politics is no politics. This situation seldom exists.

6. The literature in social experimentation on this topic is reviewed by Dennis (1988, sec. 2); the importance of the vulnerability is stressed by Orr in Harris (1985).

7. The importance of adaptive management as a challenge to the methods of scientists was first made clear to me in a conversation with Gordon Orians.

4. Gyroscope: Negotiation and Conflict

1. I am grateful to Harlan Wilson for initially recommending this book to me. Professor Wilson of course bears no responsibility for any errors in my reading of Dewey.

2. It is interesting, in this respect, that nuclear power, which had an unfavorable media image from the start of the commercial nuclear industry, dropped only slowly in public opinion polls, going below 50 percent approval only after the Chernobyl accident in the Soviet Union in 1986.

3. In the terminology used by Calabresi and Bobbitt (1978, p. 19), a sustainable fishery is one in which the "first order" determination of the size of the sustainable harvest comports with the "second order" allocation of the catch. The biological capability of the Columbia basin may make such a sustainable fishery possible.

4. Sabatier (1988) draws upon a large number of studies of policymaking and politics in American government, but makes a distinctive contribution by emphasizing the learning that takes place within policy subsystems.

5. The significance of redundancy in organizational and policy design was initially recognized by Landau (1969) and later developed systematically by Bendor (1985).

6. This discussion relies on Taylor's (1984) insightful analysis.

7. The EIS process has forced developers and agencies to make substantial improvements in project design and operating procedures; the trans-Alaska oil pipeline is a notable example. Yet there has been little cumulative learning based upon the environmental work done in the course of preparing impact statements.

8. To see why planning can be attractive early in a potential dispute, consider an analysis drawn from Chubb (1983, pp. 224–27). Planning is an activity in which matters of high value are at stake, so that a wide spectrum of interests is motivated to participate. Barriers to entry are usually low at the outset and can be kept low by planners who put a priority on doing so. Becoming an effective participant in planning does not require much previous experience: established relationships are usually weak at the beginning of the process, and there is often substantial uncertainty about how the links among different interests will be changed. Planners can exercise considerable discretion in creating and organizing the social process. When implementation is important to planners, they turn to the outside world: when they are "useful rather than simply entertaining" (Ascher and Brewer 1988, p. 216), analysts take seriously that planning is "either political or . . . decorative" (Chubb 1983, p. 224).

9. In social science jargon this is called Prisoner's Dilemma. My discussion follows that of Axelrod (1984).

10. Consensus among the affected constituencies is important to long-range remedial actions such as the Columbia basin program. Remedial actions typically encounter problems of economic justification. Damages from past actions are a sunk cost: the value of the resource has been taken by the exploiter and is no longer available to pay for remediation. But the damaged ecosystem may also contain hidden opportunities, since the ability of natural systems to recuperate is often uncertain. Under these circumstances strict cost-benefit estimates are likely to undervalue the worth of rehabilitating the ecosystem, especially if it is difficult or impossible to fund rehabilitation from the profits of exploitation.

    When these conditions occur, a negotiated consensus reflecting a mandate for rehabilitation is needed to justify expenditures and to agree on which biological risks to take. If past damages have altered the political environment by driving out a group of resource users (such as the Indians of the Columbia basin), rebuilding a sustainable suite of uses may require that points of view silenced by earlier misuse be actively sought out. Consensus-building has been strategically important in the Columbia basin, where a central agent finances decentralized actions, no one of which meets a narrow cost-benefit test, even though their cumulative impact may be economically sound.

11. Two of the most important of these forces and their interactions were examined by Hirschman (1970). "Exit" is the characteristic signal of decline in markets, as failing sellers lose buyers; "voice"—protests, appeals to authority, or political competition—is characteristically a

signal of decline in political settings. Hirschman's major point is that the human tendency to ignore bad news can be abetted by social arrangements that weaken or mask signals of decline. Hence, in markets voice is underrated, and in politics exit is ignored for too long. Countering these effects is an important but unrecognized task of planning and evaluation.

12. As a practical matter, conflicts could not be permanently resolved by the moving of boundaries. To begin with, governments and their constituencies are unlikely to reorganize jurisdictions around ecosystem boundaries. Even if this were feasible, there would be scientific difficulties. The definition of an ecosystem varies according to the analytic or management objective. For example, the Columbia drainage is too large for the management of most terrestrial wildlife, which inhabit tributary drainages; it is too small for the analysis of mixed-stock fisheries in which fish from several rivers are caught together in the ocean; and it is simply inappropriate for the management of migratory birds, whose ecosystems contain flyways overlapping parts of areas far removed from the Columbia. Even within a purely technical framework ecosystem-based management encounters "jurisdictional" puzzles. See Orians (1980).

In the nineteenth century John Wesley Powell attempted to organize the still-unsettled western United States around river basins. See Volkman and Lee (1988).

5. Sea Trials: Comparison Cases

1. It is unclear how TFW has affected timber harvest, because the world market for forest products improved sharply at the same time. The value of timber harvested annually increased by nearly 25 percent on Washington's nonfederal lands, the number of applications submitted to DNR increased by approximately 50 percent, but the actual volume of timber increased only about 5 percent. TFW increased the government's capability to regulate cutting. Small landowners, responding to an improved timber market, are believed to account for a larger fraction of logging. Although the causal relationships between increased timber value and number of applications and the TFW agreement are unclear, at the least the agreement has not thwarted the economic goals of the timber industry.

2. The epistemic community can provide insight because its shared procedures lead to reliable knowledge. That is, an epistemic community produces benefits similar to adaptive management. Holzner and Marx (1979, p. 109) trace this distinctive capability of an epistemic community to the way the community tests the claims made by its members.

3. Scoullos (1991) also takes a slightly different view of the epistemic community, arguing that an elite circle of European scientific and governmental leaders was behind both the 1972 Stockholm conference and the formation of UNEP. Members of *this* elite also took an interest in the Mediterranean and the Med Plan, laying the groundwork for the hiring of scientifically trained staff in national ministries.

Scoullos notes that the worldwide "regional seas" program of UNEP has generally been far less successful than the Med Plan, a result he attributes to the absence of the European elites in the other arenas. This elite circle derives its influence from its members' social positions in their own countries, not their joint commitment to science as a way of understanding reality; that conceptual difference is important to Haas's argument, although I do not rely on it in my own analysis of sustainability.

6. Navigational Lore: Expectations of Learning

1. Governmental activities cannot be treated as if they were simply economic transactions in another guise (Moe 1984), even though economic analysis is often helpful in considerations of public policy. Pratt and Zeckhauser (1985, p. 3) note, for example, that principal-agent theory usually assumes there are no other parties at interest besides principal and agent—that is, that there are no spillover effects. In the environmental arena this is seldom the case.

2. These mitigations can be thought of as responses to "threats to the integrity" of the (quasi-market) transactions between agent and principal. As in the case of threats to the validity of quasi-experiments, discussed in Chapter 3, the issue is to assess and to mitigate, as well as to clarify transactions that should be avoided.

3. More generally, enforcement can work even if it is infrequent, so long as penalties when one is caught are high.

4. Strictly speaking, however, this shared incentive exists only at the outset of a relationship, because subsequent changes amount to changes in already-agreed-to contracts. Accordingly, changes in midcourse may require grandfather clauses, side payments, or other mechanisms to maintain equity (Pratt and Zeckhauser 1985, pp. 19–20).

5. This is an alternative explanation of research and development by monopolistic firms. The classic explanation of Schumpeter (1942) was based on property rights: a monopolist could be confident of his ability to keep the benefits of research. Levinthal and March (1981, p. 200) suggest an explanation less directly linked to purpose: those with market power are successful and will tend to invest in learning, including research, for reasons that may or may not have to do with eventual repayment.

6. A market accomplishes complex coordination without having or needing a central purpose. One of the strengths of market economies is precisely that they avoid the need to have detailed economic objectives. In doing so, however, they sacrifice many other things, notably the possibility of pursuing goals such as social justice.

7. See Lustick (1980). This way of thinking about budgets in large organizations originated with Cyert and March (1963).

8. An interestingly complex example is Taylor's (1984) analysis of the implementation of the National Environmental Policy Act.

9. The terminology of single- and double-loop learning was invented by Argyris and Schon (1978).

10. Andrews (1990) suggests that ideological presumptions can exert a strong bias in risk analysis, one of the principal technical tools used to develop environmental policy. In the context of this discussion of theories of learning, ideology is a source of bias, a way of rationing the scarce attention that individuals or organizations focus on problems. Andrews's critique points toward a much larger debate, the ideological values of environmentalism and its rivals.

7. Seaworthiness: Civic Science

1. See Price (1965, chap. 5). The adaptation in Fig. 7–1 was developed in discussions over several years with Todd R. LaPorte and Gordon Orians.

2. Ernst Haas (1990, p. 171) puts it this way: "Whenever the leadership of an organization heavily dependent on scientific and technical personnel fears that knowledge has changed so as to put into question the mission of their organization, the scientific component risks being corrupted because scientific questioning is choked off. The original knowledge then becomes dogma."

3. The political cost is much lower when the cause is external to the sponsoring agency—such as a cutback forced by economic recession and sharply lower budgets. But in this case, the evaluations cannot be done in time on an experimental basis, so the cutbacks would be made on less reliable knowledge. This is an example of institutional constraint acting as a cognitive bias.

4. This discussion develops ideas articulated in Selznick (1957) and Taylor (1984). The principal difference lies in my focus on networks of organizations, of the kind implicitly treated by Taylor, rather than on the single organization analyzed by Selznick.

5. See Taylor (1984, pp. 258–74).

6. For a helpful historical perspective on the role of visualization in the evolution of technology, see Ferguson (1977).

7. "Doing business" includes consumption as well as production. It is central to the pursuit of sustainability that we gather a coherent picture of changes due to both consumptive activities and production.
8. In a useful article, Alvin Weinberg (1972) described such situations, in which scientific questions could not be answered with purely scientific responses, as ones in which "trans-science" judgments were necessary.

8. Seeking Sustainability

1. This approach—to identify boundary conditions that must be satisfied by any solution to a problem, before setting out to find the solution explicitly—is borrowed from mathematics. In environmental policy there are no general principles of sufficient precision to develop explicit solutions, so one must proceed empirically, from experience. But having knowledge of the boundary conditions narrows the range of experience needed to illuminate promising paths.
2. There are, of course, many other dimensions to the relationship between people and the natural world. When an economic transaction treats as a commodity something that is functionally integral to an ecosystem or laden with meaning to humans, conflict is often the result. In this the struggle of the Penan of Sarawak to save their rain forest is akin to the defiance of loggers in the Pacific Northwest intent on the harvest of old-growth forests. Both seek to secure a way of life against the battering of markets.
3. A contemporary commentator puts it this way: "Lost on us is the cruelty of that cruelest month, April, when what one had stored of last year's harvest would be running out, while nature greened with promise but couldn't be expected to deliver until its due season" (Gifford 1991, p. 124).
4. International trade required something akin to money—a compact store of value that could be readily exchanged and transported (see Cronon 1983, chap. 5). The expansion of a money economy made it possible to accumulate wealth—something that nomadic life did not often permit.
5. This is an area marked by lively controversy. A useful summary of the debate is Turner (1988).
6. A noneconomist finds it surprising that economic theorists have considered sustainability only within the framework of national economies, although it is apparent that international trade has a substantial influence on exploitation of natural resources everywhere. Others who have puzzled over the definition of sustainability include Brown et al. (1987), Liverman et al. (1988), and Worldwatch Institute (1987). The latter two sources are particularly rich in data; the clearest exposition of

concepts is Tietenberg (1992), whose analysis I rely on at several points. The view developed here is compatible with that advanced in Redcliff (1987, chap. 2), although my discussion is limited largely to neoclassical economic ideas.

7. Rates of return taking into account the erosive power of inflation—called "real" rates of return by economists—have been negative for some periods.

8. The discount is the rate of investment at a risk equal to that of the project under consideration and with the effects of inflation removed. Historical data, however, record market rates, which combine risk and inflation. Inflation can be removed in a fairly straightforward fashion, but risk is more difficult to take into account. I am grateful to Stephen DeCanio for an enlightening discussion of the complexities of estimating discount rates from historical data.

9. This range was provided by Stephen DeCanio, with the caution that much of the capital in the economy is in housing, a form of investment in which the rate of interest is not directly observable, so that even the wide range cited may not be wide enough. The current practice of requiring 10 percent return is taken from Committee on Science, Engineering, and Public Policy (1991, chap. 2).

10. The logic of discounting operates across an economy, not just in specific markets. Although the discussion in the text suggests that a lower discount rate would be beneficial in the sense that it would justify slower exploitation of natural resources, this is not necessarily the result when the economy as a whole is taken into account. A lower discount rate would also encourage more investment in aggregate, because lower discounts cause future benefits to loom larger, so that people find it advantageous to shift resources from consumption to investment. More investment can easily increase the flow of material goods, and environmental damage along with it (Pearce, Barbier, and Markandya 1988, pp. 23–24).

11. There is no settled view on whether all species in an ecosystem must be preserved if the ecosystem is to be managed in a sustainable fashion. This is partly a matter of ecological uncertainty—we do not know the conditions under which all species are functionally necessary for a managed ecosystem to be stable—and partly a matter of philosophical principle.

12. An extraordinary example is the deliberate elimination of guns as instruments of war in feudal Japan (Perrin 1979)—something that may have contributed to the military inability of its rulers to resist the pressures of the American navy to open the nation in 1858.

This argument assumes two patterns: that military power is proportional to economic output, and that national welfare requires political independence that must be defended militarily. Postwar Japan and, to a lesser degree, Western Europe have operated as exceptions to these general patterns. Neither the postwar superpower rivalry nor whatever is now emerging from its collapse is likely to be a clear guide to the long-term future.

Different analyses are clearly possible, however. Stephen DeCanio has suggested that future generations will want the current generation to conserve environmental values because people in the future will be richer and able to spend more on environmental protection. If they were able to do so, they would transfer some of their wealth back to us, to pay for enduring environmental protection today. (DeCanio's argument contributed to the U.S. position on the 1987 Montreal Protocol on Substances That Deplete the Ozone Layer.) This argument is similar to the idea of limiting consumption mentioned in the text and described in Tietenberg (1992, pp. 619–21).

13. The French philosopher Blaise Pascal suggested a rational argument for believing in God. If one believes and there is a God, one achieves eternal salvation; if one believes but there is no God, one has lived a virtuous life. If one does not believe, the pleasures of earthly dissipation cannot outweigh the risk of eternal damnation.

14. The division into industrial democracies (First World), socialist command economies (Second World), and developing countries (Third World) is part of the language of the cold war.

15. Heclo's comment on the historical process of social learning is apt: "social policy is not like a shoe or a loaf of bread; it is too complex to be explained simply as the predicate of some 'maker.' We should seek to examine not only who has contributed to social policy but how their contributions have been related. We should inquire not only how things work but how, if at all, the working of things has changed through time" (1974, p. 9).

Notes on Sources

The notes below provide sources for direct quotations, acknowledge intellectual debts, and indicate sources of further reading.

Preface

The quotations are from United Nations World Commission on Environment and Development (1987, p. 22) and Heilbroner (1974, p. 132).

Prologue: After Columbus

These pages draw upon Brooks (1977), Heclo (1974), and Wallace Stegner's essay "Inheritance" (1983). Stegner's biography of John Wesley Powell (1954) demonstrates the ability of a single person to shape a nation's thinking about its land—and the frailty of having only a single person do so. The power of humans as an environmental force is discussed by Worster (1985), Clark (1989), the National Research Council's Committee on Science, Engineering and Public Policy (1991), and, with historical depth and lucid breadth, by Cronon (1983). A skeptical perspective about that power is offered in Gould (1990), and a commentary on how far we have come in a brief time is given by Gifford (1991, pp. 219–20). Mathews (1989) observes that the poor of our planet remain desperately vulnerable to the quality of the environment from which they gain their livelihood. The concept of a "land ethic" was advanced in the 1940s by Leopold (1949, pt. IV).

Chapter 1. Taking Measures

Kennan reflected on the environment in his memoirs (1972, vol. 2, p. 85); the words of Billy Frank, Jr., are taken from Egan (1992). The now-standard definition of sustainable development ("development that meets the needs of the present without compromising the ability of future generations to meet their own needs") is taken from the United Nations Environment Programme (1989). On possible conceptual interpretations of the term, see Tietenberg (1992, chap. 22) and Redclift (1987).

Our Common Future, the report of the United Nations World Commission on Environment and Development (1987), sets the terms of the debate which this book joins. The facts on humans' role in the environment are drawn from Clark (1989). Adaptive management, discussed in detail in Chapter 3, was formulated initially by Holling (1978); I first encountered the notion of bounded conflict in Coser (1956) and Truman (1971). My understanding of "ecosystem" as an idea that humans use to organize natural phenomena is drawn from Orians (1980). How organizations and networks of organizations respond to complex opportunities such as the large ecosystem is discussed suggestively by March and Olsen (1984). The situation of developing countries is summarized briefly by Silk (1989) and conceptualized helpfully by Redclift (1987). Nectoux and Kuroda (1990) and Sesser (1991) provide a useful perspective on the rain forest of Borneo. Cronon's (1983) insightful interpretation of Native Americans as participants in the ecosystem of New England demonstrated to me how complex sustainable development must be in the presence of international trade. Corroborating evidence is provided in Trigger (1991).

Chapter 2. Sustainability in the Columbia Basin

Hughes's appreciation of technology as a social force is quoted (1983, p. 1), together with words from Egan's (1992) profile of Billy Frank, Jr. Comparisons between the Columbia and Colorado are drawn from Kahrl (1978, p. 3). The population estimate for Native Americans in the Columbia basin is taken from Schalk (1986). Harold Culpus is quoted in *Northwest Energy News* 1, no. 3 (May/June 1982), 10–11. Lines from three songs by Woody Guthrie (1941) are quoted in the text. For a more detailed analysis of the matters discussed in this chapter, with fuller references to the literature, see Lee (1991a, 1991b).

The best introduction to the problems of the salmon in the Columbia basin is by Wilkinson and Conner (1983). Developments since then have been described in *Northwest Energy News*, the newsletter and later magazine of the Northwest Power Planning Council—as vivid a demonstration as Thomas Jefferson could have wished that public involvement can be interesting, even-handed, and educational. The figures in the text are based on data on the salmon populations assembled by the council. Analysis from a legal (and often polemical) perspective has come from Professor Michael C. Blumm of the law faculty at Lewis and Clark. His articles, appearing mostly in the Lewis and Clark law review, *Environmental Law*, are complemented by more detailed commentary in the *Anadromous Fish Law Memo*, published by Oregon Sea Grant, Corvallis, which he edited until 1990; an overview is provided in Blumm and Simrin (1991). Whether one agrees with Blumm or not (see Lee 1991b), his published works provide an indispensable continuity. Blumm's view on the standard to be met in reallocating river flow to benefit fish (1983, p. 129) is quoted in the text. The

Northwest Power Planning Council (1987a, app. E) provides additional historical information and a careful estimate of losses of salmon (summarized at pp. 37–38). The Columbia River Inter-Tribal Fish Commission (1982), Lothrop (1986), and Platt and Dompier (1990) describe how fisheries and hatcheries have been buffeted by politics; and Cohen (1986) describes the crucial Indian treaty litigation that shifted rights to harvest salmon. The problems of hatcheries as a replacement for the natural habitat have been ably summarized by Goodman (1990), and in a polemical style by Brown (1982). Lee (1989), Muckleston (1990), and Volkman (1992) link Columbia basin policy to sustainable development.

Worster's judgment of the fate of Western rivers is quoted (1985, p. 276). Two earlier, more optimistic appraisals of the possibilities of multiple use in river basins are McKinley (1952) and the influential memoir of Lilienthal (1943). The views of environmentalists tend to be harsher (see Reisner 1986). All these writers chronicle a "tragedy of the commons," a term added to the lexicon of the environment by Hardin (1968).

The story of the Bonneville Power Administration and its regional influence is chronicled by McKinley (1952); Lee, Klemka, and Marts (1980); and Tollefson (1987). All seek to interpret the information, both eloquent and abstruse, in the annual reports of the agency (e.g., U.S. Department of Energy 1989b), which tell Bonneville's story in dry but sometimes dramatic terms; the 1989 report describes the dollar costs of the salmon rehabilitation effort as well as the increase in rates in the early 1980s. The desperate story of the Washington Public Power Supply System, a minor subplot here, is developed by Chasan (1985), Sugai (1987), and Leigland and Lamb (1986). Hughes (1983, pp. 363–64) discusses the long, profitable decline in the cost of making electricity.

The Northwest Power Planning Council (1987a, 1991) articulates the innovations in policy mandated by the Northwest Power Act (U.S. Congress 1980), notably the doctrine of cost-effectiveness and the systemwide approach to the Columbia basin. The vision of the council's founding members is set out in Evans and Hemmingway (1984); Hemmingway's (1983) interpretation of the council as an interstate compact proved sensible to the federal judiciary (see Seattle Master Builders Association et al. 1986); and founding member Charles Collins is quoted at the end of the chapter in Mahar and Riley (1985). Lee (1983) discusses the council's risk-managing strategy, a prototype of least-cost planning, the approach now widely favored by utility regulators. Volkman and Lee (1988) discuss the applicability of the institutional model of the council elsewhere in the United States. I have appropriated Wildavsky's definition of majority rule (1979, p. 259).

Although energy efficiency is now a well-accepted concept, it was not always so; the observations and critiques of Eric Hirst of the Oak Ridge

National Laboratory have played an important role in making conservation real to managers and engineers in the utilities. An example is the evaluation of the Hood River Project by Hirst, Goeltz, and Trumble (1987). Figures on the costs and energy saved at Hood River are found in Philips et al. (1987, p. 48). The challenges of conservation have been candidly discussed by Bonneville (see U.S. Department of Energy 1990), the U.S. General Accounting Office (1987), the Northwest Power Planning Council (1989), and, from a national perspective, by Ruderman, Levine, and McMahon (1987). Koomey's (1990) dissertation, decidedly favorable toward conservation, provides a good summary of the literature of a field that sprawls across engineering, economics, marketing, and sociology. The problem of "free riders," those who benefit from conservation even when they do not participate, was originally formulated in general terms by Olson (1965).

The system planning language (section 200) of the fish and wildlife program is particularly germane to the discussion in this book. Lawrence (1983) has documented the development of the water budget. I have relied on estimates of the cost of the water budget and other changes in river operations by James D. Ruff of the planning council staff. My discussion of the water budget's biological effectiveness relies upon information assembled by Sims and Ossiander (1981). The quoted passage on burden of proof is from Blumm (1982, p. 129).

Chapter 3. Compass: Adaptive Management

The epigraph to the chapter is from Sorensen (1963, p. 21). The initial illustration of fish harvest regulation as an opportunity for experimentation is based upon Argue et al. (1983). Leopold (1949, p. 240) provides the dictum on the proper role of *Homo sapiens*. The injunction to discover solutions is from Landau and Stout (1979, p. 154). That environmental quality cannot be achieved by eliminating change is argued by Holling (1978, p. 33). The sad story of the Kemp's Ridley sea turtle is summarized by Taubes (1992). Newton's delay in publishing the *Principia Mathematica* is discussed clearly by Park (1988, pp. 183–85). The quotation contrasting trapped with experimental administrators is from Campbell (1969, p. 200). Harry Wagner's musings about the problems of rebuilding salmon runs in "Pandora's basin" are from Collette (1992, p. 30), and doubt about whether the Columbia basin can persevere in learning is expressed in Volkman (1992, p. 39).

Bounded rationality is an idea developed by Herbert Simon (1954, 1983) and his students, notably James March, who emphasized in a path-breaking book (Cyert and March 1963) that information—and therefore learning—is costly to acquire.

This chapter draws upon three principal sources. The headwaters are

Holling's original formulation of adaptive management (1978); the similar approach of holistic resource management is discussed by Savory (1988). Ideas about experimentation in social settings, particularly the concepts of internal and external validity, derive from Campbell (1969). LaPorte (e.g., LaPorte and Consolini 1991), Miles (1987), and Landau (1969) taught me how to think about organizations in a technical and scientific context. Walters (1986) codifies the work of the Holling group by developing the technical framework of ecosystem modeling. Hilborn (1987), another of Holling's students, provides the basis for explaining why knowledge of ecosystems is sparse and incomplete. The broader question of how ecological science is related to environmental policy has been framed authoritatively by the National Research Council (1986).

The early experience of adaptive management is discussed by Environmental and Social Systems Analysts, Ltd. (1982). Application of adaptive management to the Columbia basin is discussed in Lee and Lawrence (1986) and articulated in Northwest Power Planning Council (1987a, sec. 200). Miles (1987) makes the point that research that has clearly foreseeable consequences for policy will tempt policy actors to meddle—a concern that strikes at the heart of whether adaptive management is possible in real life. Though not framed in this fashion, much of Chapters 3 and 4 endeavors to answer Miles's difficult and essential skepticism.

My discussion of "avoidable error" relies upon Campbell's original analysis, including the quotation on the need for quasi-experimentation (1969, p. 195). The discussion of experimentation also relies upon the able summary in Dennis (1988); Boruch, Dennis, and Carter-Greer (1988); Lempert and Vissher (1988); and Achen (1986). The Hawthorne effect is summarized by Scott (1987, pp. 57–58). Harris (1985) contributed to the literature of experimentation the economist's awareness of how general equilibrium can differ from short-run response, as those affected by a policy react to its imposition. Gordon Orians introduced me to Type I and II errors, power of test, and the novelty of ecosystem-scale experimentation for university-trained ecologists; Peterman (1990) provides a clear summary for someone like me who has only a superficial acquaintance with statistics.

Chapter 4. Gyroscope: Negotiation and Conflict

Holling's comment on boundaries (1978, p. 11) begins the chapter. Dewey's hopes for experimentation (1927, p. 34) and his words concerning the need for better debate (1927, p. 208) are also quoted. Frederick Douglass is quoted in Cormick (1980, p. 25). The explanation of tragic choice is based on passages from Calabresi and Bobbitt (1978, p. 18). The characterization of the loss of species as incalculable is taken from a decision of the U.S. Supreme Court (*Tennessee Valley Authority v. Hill 178* [1973]). The wry

definition of "majority" rule is from Wildavsky (1979, p. 259). The consti-
tutionality of the Power Planning Council was clarified in Seattle Master
Builders Association et al. (1986). I quote Taylor's evaluation of the
courts' ability to keep agencies on their toes (1984, p. 90) and Kingdon
(1984) on the importance of ideas in policy debate (p. 131), the biochemical metaphor (pp. 122–23), and the image of ideas suddenly taking
off (p. 85).

Gerald Cormick taught me to appreciate conflict, supplemented by
Coser (1956), Nader and Todd (1978) and Shapiro (1975). Hardin (1968),
Amy (1987), Worster (1985), and Hays (1987) describe the logic and history
of environmental conflict, while Baskerville's (1988) dictum that planning
centralizes while implementation decentralizes has shaped my thinking
about how learning occurs and where it needs to be protected.

The deep connections among conflict, social error correction, and political stability are a principal theme of American political thought in the
twentieth century, exemplified here in Dewey (1927) and later in Truman
(1951). Ways in which the self-regulating capability of political competition may be undermined are discussed by Downs (1972) and Calabresi and
Bobbitt (1978). Sabatier (1988), Taylor (1984), and Kingdon (1984) suggest, however, that learning in public policy occurs more often than might
be expected, although the process is often perverse (Landy, Roberts, and
Thomas 1990) or accidental (Kingdon 1984) rather than logical. When
politics has a random element, how institutions steer people's attention
becomes surprisingly important, a point that March and Olsen (1989)
develop with many insights. Accordingly, redundant mechanisms (Landau 1969, Bendor 1985) can provide methods for keeping lines of communication open (Taylor 1984). Negotiation among competing parties is
another means of attaining this end; Amy's (1987) discussion is persuasive.
My way of telling the story relies on Thompson and Tuden (1959), follows
Baur's (1977) emphasis on the intervener's role in organizing the dispute,
and interprets the political tensions of planning as suggested by Chubb
(1983, pp. 224–27) and Hirschman (1970). Putnam (1988) comments helpfully on the relationship between leaders who negotiate and followers
whose support is needed by leaders. Axelrod's (1984) influential discussion
of how cooperation can be built *without* central authority has led me to be
more inventive about institutional arrangement. Regier and Baskerville
(1986) provide a thoughtful perspective on the dilemma of ecosystem
rehabilitation. Reading the minutes of one of the Northwest Power Planning Council's consultative committees (1988–1990), I could see error
correction at work, as utility representatives with no interest in biology
challenged fisheries interests with little concern for economics, producing
the sort of vigilance that is essential to the effectiveness and sustainability
of the adaptive process.

Chapter 5. Sea Trials: Comparison Cases

The section on the Queensland tropical forest relies on Poore et al. (1989, chap. 2) and on Keto and Scott (1986) and Australian Ministry of the Arts, Sports, the Environment, Tourism and Territories (1988). I am indebted to Mara Bún for assistance in obtaining the Australian sources. I have excerpted Harvard biologist Peter Ashton's doubts about Queensland's research program, as quoted in Keto and Scott (1986). The long quotation is from Poore et al. (1989, p. 238).

The section on the Timber, Fish, and Wildlife Agreement is adapted from Halbert and Lee (1990). The term *forest practices* includes harvesting, logging, roadbuilding, reforestation, and application of chemicals to forest lands. Language from the original agreement is taken from TFW Final Agreement (1987). Amy's point about the advantages of pooling information is quoted (1987, p. 37).

Ray Hilborn (1992), co-investigator with me on a three-year-study of adaptive management, did the empirical work on the sockeye spawning channels.

The discussion of social experimentation is drawn from Rivlin (1971); Campbell (1969); Dennis (1988); Boruch, Dennis, and Carter-Greer (1988); and Lempert and Vissher (1988); the last three sources review the uneven history of social experimentation. The gloomy assessment is quoted from Boruch, Dennis, and Carter-Greer (1988, p. 412). The warning about quasi-experiments is from Achen (1986, p. 161).

Peter M. Haas's innovative study (1990) grounds my discussion of the Med Plan, supplemented by Batisse (1990) and a helpful discussion with Dr. Michael Scoullos (1991), arranged for me by Professor Toshiko Akiyama. Quotes are taken from P. Haas (1990, pp. 27–28, 55, 78, 131–32, 227, 229).

The brief treatment of learning in multilateral institutions borrows from Ernst Haas (1990), which surveys a wide literature with his customary power. Quoted passages are from E. Haas (1990, pp. 26, 27, 30, 72, 122, 193).

Chapter 6. Navigational Lore: Expectations of Learning

The epigraph is from Dewey (1916, p. 216).

The theoretical framework I propose here owes much to Levitt and March (1988) and Heimer (1988) and to two elegant lines of research pursued largely at Stanford during the 1970s: on organizations and choice, by James March and his students (see March 1988); and on inherent biases in human cognition, by Daniel Kahneman and Amos Tversky (see Kahneman, Slovic, and Tversky 1982).

Rational learning is discussed by Steinbruner (1974) as a baseline for exploring cybernetic learning. I have appended the useful institutional analysis by Woodhouse (1988) of risk management, a term propounded by Ruckelshaus (1985). The account of principal-agent tensions relies on summaries by Pratt and Zeckhauser (1985), Moe (1984), and Bendor (1988). Wolf's (1979) skeptical appraisal of governments criticizes the tendency to assume that governments do what they are supposed to do; that approach is helpful in thinking about principal-agent difficulties.

The topic of bounded rationality originated with Herbert Simon. Steinbruner (1974) developed the notion of cybernetic learning; Heclo (1974) did much to show how bounded rationality influences our understanding of large-scale public policies. Building on Simon, March has insightfully decoupled learning from reason in the notion of superstitious learning (March and Olsen 1976, chap. 4; Levitt and March 1988); earlier, March proposed that bounded rationality in economic organizations could be understood in political terms (Cyert and March 1963), so that concepts such as coalition and negotiation, derived from the study of politics, could be applied to corporations. The Simon-March literature has tended to stress the need to expand the idea of rationality to describe successful entities with multiple minds and therefore no clear priorities. (Kingdon's organized anarchy is the intellectual descendant of this line of inquiry.) Argyris and Schon (1978) have taken bounded rationality in a darker direction, arguing that limited understanding can create crisis when learning is constrained to be single-loop.

The topic of cognitive and institutional bias has been shaped, for me, by the insights won by Kahneman and Tversky. Helpful perspective on the degree to which we should mistrust everyday experience may be found in Edwards and von Winterfeldt (1986). That there are inherent limitations on the accuracy of human perception echoes Hirschman's wise recognition (1970) that our methods of detecting and correcting error are themselves subject to breakdown. The notion that goals are often ambiguous is one I learned originally from March (1972). March's work leading up to Kingdon's analysis of organized anarchy (1984) shapes the way I have tried to carry the ideas of cognitive bias into an institutional setting. Heimer (1988) suggests that redundancies such as those discussed in Chapter 4 are a means of overcoming or mitigating these inherent limits. Andrews (1990) suggests that ideological presumptions can exert a strong bias in risk analysis. Western water law (Wilkinson 1989, Worster 1985) provides an instructive parallel to cognitive bias in individuals; the definition of property claims on flowing water preempts environmental considerations that are important in this century but were irrelevant to the rush toward Manifest Destiny in the nineteenth. What I have called "patient" learning is drawn from Levitt and March (1988).

Chapter 7. Seaworthiness: Civic Science

I start with a quotation from Dewey (1927, p. 225). Ray Hilborn (1990) provided the example of an indirect method of estimating the cost-effectiveness of hatcheries in the Columbia. The story of the model that assumed the Columbia River would flow uphill when needed comes from the Pacific Northwest Utilities Conference Committee (1977). The quotation describing social learning is from Heclo (1974, p. 308). Cost estimates for the Yakima hatchery are taken from the Northwest Power Planning Council (1990, table 12, p. 41).

Brewer's (1973) clear thinking about how public policies should be judged has influenced my discussion in this chapter. My proposal that the broad term *civic science* be analyzed along a small number of criteria adopts Brewer's concept of appraisal. The conceptual ground for thinking about science and politics was defined with insight by Price (1965). Their tension is exemplified by the Faustian bargains that trade off scientific rigor for political independence—the tradeoff that Miles (1987) suggested would be hard to avoid. The tension can be shifted from the organizational to the individual level, if there is a group of professionals willing to describe the hard choices as ones needing their expertise (Selznick 1957).

The discussion of Type II errors can be explored further in Peterman (1990). Landy, Roberts, and Thomas (1990) describe the difficulty faced by regulators in setting clear standards for those releasing pollutants into the environment. The discussion of visualization takes Ferguson (1977) as its point of departure.

Chapter 8. Seeking Sustainability

The opening quotations are from Whitehead (1925, p. 98) and Aung San Suu Kyi (1991, p. 183), whose politically courageous essays have won her a Nobel prize and imprisonment. That the meaning of affluence is shaped by social circumstances has been developed by Sahlins (1972), whose analysis I have foreshortened considerably in a way suggested by Cronon (1983, pp. 79–80). The cost to the Ethiopian economy of burning dung was estimated by Newcombe (cited in Pearce 1988, p. 105). The image of sacrifice at the altar of beliefs is from Berger (1974, p. 3).

History reaches back to times of biological significance. Economics, because it is naturally framed in quantitative terms, permits extrapolation across ranges of time and human welfare that correspond to dramatically changed relations between people and environment. As must be the case with any conceptual treatment of sustainable development, history and economics are the principal bases of this chapter. I first encountered the heroic dimensions of this discussion in Heilbroner (1974); Sahlins (1972) provides a strikingly different perspective on poverty as a culturally mediated perception. Here I have used William Cronon's lucid discussion of New England in the early colonial period (1983) to stand for the wider, far more complex conquest of the known world by European culture. The World Bank's annual *World Development Report* has discussed sustainability as a policy objective with authority and lucidity, particularly in the 1992 report. Basalla (1988) has synthesized the history of technology in a way that sheds light on the possibilities for the continuance of economic growth into the future. The economic concept of the discount rate, which permits comparison of values across long periods (at the cost of heroic assumptions), is the tool I use for exploring the future, following the excellent discussion in Tietenberg (1992, chap. 22); conversations with Professors Gardner Brown, Stephen DeCanio, and Tom Tietenberg have been helpful in shoring up my limited appreciation of the complexities of economic thought in this often-plowed field.

Bibliography

Achen, Christopher H. 1986. *The Statistical Analysis of Quasi-Experiments*. Berkeley: University of California Press.

Amy, Douglas J. 1987. *The Politics of Environmental Mediation*. New York: Columbia University Press.

Andrews, Richard N. L. 1990. "Risk Assessment: Regulation and Beyond." In *Environmental Policy in the 1990s*, ed. Norman J. Vig and Michael E. Kraft. Washington, D.C.: CQ Press.

Argue, A. W., et al. 1983. "Strait of Georgia Chinook and Coho Fishery." *Canadian Bulletin of Fisheries and Aquatic Sciences*, no. 211 (whole issue).

Argyris, Chris, and Donald A. Schon. 1978. *Organizational Learning: A Theory of Action Perspective*. Reading, Mass.: Addison-Wesley.

Ascher, William, and Garry D. Brewer. 1988. "Sustainable Development and Natural Resource Forecasting." In *Redirecting the RPA*, ed. Clark S. Binkley, Garry D. Brewer, and Alaric V. Sample. Yale School of Forestry and Environmental Studies, bull. 95, 216–29.

Aung San Suu Kyi. 1991. *Freedom from Fear, and Other Writings*. New York: Penguin.

Australian Ministry of the Arts, Sports, the Environment, Tourism and Territories. 1988. "World Heritage Protection for the Wet Tropical Rainforests." *Ecofile*, special issue, August.

Axelrod, Robert. 1984. *The Evolution of Cooperation*. New York: Basic Books.

Basalla, George. 1988. *The Evolution of Technology*. Cambridge: Cambridge University Press.

Baskerville, G. L. 1988. "Redevelopment of a Degrading Forest System." *Ambio* 17:314–22.

Batisse, Michel. 1990. "Probing the Future of the Mediterranean Basin." *Environment* 32, no. 5 (June): 5–9, 28–34.

Baur, E. Jackson. 1977. "Mediating Environmental Disputes." *Western Sociological Review* 8:16–24.

Bendor, Jonathan B. 1985. *Parallel Systems*. Berkeley: University of California Press.

————. 1988. "Review Article: Formal Models of Bureaucracy." *British Journal of Political Science* 18:353–95. Republished as "Formal Models of Bureaucracy: A Review," in *Public Administration: The State of the Discipline*, ed. Naomi B. Lynn and Aaron Wildavsky. Chatham, N.J.: Chatham House Publishers, 1990.

Berger, James O., and Donald A. Berry. 1988. "Statistical Analysis and the Illusion of Objectivity." *American Scientist* 76:159–65.

Berger, Peter L. 1974. *Pyramids of Sacrifice*. Garden City, N.Y.: Anchor.

Blumm, Michael C. 1982. "Fulfilling the Parity Promise: A Perspective on Scientific Proof, Economic Cost, and Indian Treaty Rights in the Approval of the Columbia Basin Fish and Wildlife Program." *Environmental Law* 13:103–59.

Blumm, Michael C., and Andy Simrin. 1991. "The Unraveling of the Parity Promise: Hydropower, Salmon, and Endangered Species in the Columbia Basin." *Environmental Law* 21:657–744.

Boruch, Robert F., Michael Dennis, and Kim Carter-Greer. 1988. "Lessons from the Rockefeller Foundation's Experiments on the Minority Female Single Parent Program." *Evaluation Review* 12:396–426.

Brewer, Garry D. 1973. *Politicians, Bureaucrats, and the Consultant: A Critique of Urban Problem Solving*. New York: Basic Books.

Brooks, H. 1977. "Potentials and Limitations of Societal Response to Long-Term Environmental Threats." In *Global Chemical Cycles and Their Alterations by Man*, ed. W. Stumm. Berlin: Dahlem Konferenzen.

Brown, Becky J., et al. 1987. "Global Sustainability: Toward Definition." *Environmental Management* 11:713–19.

Brown, Bruce. 1982. *Mountain in the Clouds: A Search for the Wild Salmon*. New York: Simon & Schuster.

Calabresi, Guido, and Philip Bobbitt. 1978. *Tragic Choices*. New York: W. W. Norton.

Campbell, Donald T. 1969. "Reforms as Experiments." *American Psychologist* 24:409–29. Reprinted in *Readings in Evaluation Research*, ed. Francis G. Caro, 2d ed. New York: Russell Sage Foundation, 1977, pp. 172–204, from which the page citations are taken.

Carpenter, Jan. 1988. Review of *Our Common Future*. *Northwest Environmental Journal* 4:171–72.

Chasan, Daniel Jack. 1985. *The Fall of the House of WPPSS*. Seattle: Sasquatch Publishing.

Chubb, John E. 1983. *Interest Groups and the Bureaucracy*. Stanford: Stanford University Press.

Clark, William C. 1989. "Managing Planet Earth." *Scientific American* 261, no. 3 (September): 47–54.

Clark, W. C., and R. E. Munn, eds. 1986. *Sustainable Development of the Biosphere*. Cambridge: Cambridge University Press.

Cohen, Fay G. 1986. *Treaties on Trial*. Seattle: University of Washington Press.

Collette, Carlotta. 1992. "Pandora's Basin." *Northwest Energy News*, January–February, 27–30.

Columbia River Inter-Tribal Fish Commission. 1982. "Mitigation of Anadromous Fish Losses: Efforts Related to Columbia and Snake River Dams and a Plan for Reprogramming Hatcheries." Portland, Ore.

Committee on Science, Engineering, and Public Policy, National Research Council. 1991. *Policy Implications of Greenhouse Warming: Synthesis Panel.* Washington, D.C.: National Academy Press.

Cormick, Gerald W. 1980. "The 'Theory' and Practice of Environmental Mediation." *Environmental Professional* 2:24–33.

Coser, Lewis A. 1956. *The Functions of Social Conflict.* Glencoe, Ill.: Free Press.

Cronon, William. 1983. *Changes in the Land: Indians, Colonists, and the Ecology of New England.* New York: Hill and Wang.

———. 1990. "Modes of Prophecy and Production: Placing Nature in History." *Journal of American History* 76:1122–31.

Cyert, Richard M., and James G. March. 1963. *A Behavioral Theory of the Firm.* Englewood Cliffs, N.J.: Prentice-Hall.

Dennis, Michael Louis. 1988. "Implementing Randomized Field Experiments: An Analysis of Criminal and Civil Justice Research." Ph.D. diss., Northwestern University.

Dewey, John. 1916. *Democracy and Education: An Introduction to the Philosophy of Education.* Reprint, New York: Free Press, 1966.

———. 1927. *The Public and Its Problems: An Essay in Political Inquiry.* Based on the Larwill Foundation Lectures, Kenyon College, January 1926. Reprint, Chicago: Gateway Books, 1946.

Downs, Anthony. 1972. "Up and Down with Ecology." *Public Interest* 28:38.

Edwards, Ward, and Detlof von Winterfeldt. 1986. "On Cognitive Illusions and Their Implications." *Southern California Law Review* 59:401–51.

Egan, Timothy. 1992. "Indians and Salmon: Making Nature Whole." *New York Times*, November 26, C1, C10.

Environmental and Social Systems Analysts, Ltd. 1982. *Review and Evaluation of Adaptive Environmental Assessment and Management.* Ottawa: Environment Canada.

Evans, Daniel J., and Leroy H. Hemmingway. 1984. "Northwest Power Planning: Origins and Strategies." *Northwest Environmental Journal* 1:1–22.

Ferguson, Eugene S. 1977. "The Mind's Eye: Nonverbal Thought in Technology." *Science* 197:827–36.

Gifford, Don. 1991. *The Farther Shore: A Natural History of Perception, 1798–1984.* New York: Vintage.

Goodman, Michael L. 1990. "Preserving the Genetic Diversity of Salmonid Stocks: A Call for Federal Regulation of Hatchery Programs." *Environmental Law* 20:111–66.

Gould, Stephen Jay. 1990. "The Golden Rule—A Proper Scale for Our Environmental Crisis." *Natural History*, September, 24–30.

Guthrie, Woody. 1941. *Columbia River Collection*. Reissue, Cambridge: Rounder Records, 1987.

Haas, Ernst B. 1990. *When Knowledge Is Power: Three Models of Change in International Organizations*. Berkeley: University of California Press.

Haas, Peter M. 1990. *Saving the Mediterranean: The Politics of International Environmental Cooperation*. New York: Columbia University Press.

Halbert, Cindy L., and Kai N. Lee. 1990. "The Timber, Fish, and Wildlife Agreement: Implementing Alternative Dispute Resolution in Washington State." *Northwest Environmental Journal* 6:139–75.

Hardin, Garrett. 1968. "The Tragedy of the Commons." *Science* 162 (13 December):1243–48.

Harris, Jeffrey E. 1985. "Macroexperiments versus Microexperiments for Health Policy." In *Social Experimentation*, ed. Jerry A. Hausman and David A. Wise. With comments by Paul B. Ginsburg and Lawrence L. Orr. Chicago: University of Chicago Press.

Hays, Samuel P., with Barbara D. Hays. 1987. *Beauty, Health, and Permanence: Environmental Politics in the United States, 1955–1985*. Cambridge: Cambridge University Press.

Heclo, Hugh. 1974. *Modern Social Politics in Britain and Sweden*. New Haven: Yale University Press.

Heilbroner, Robert L. 1974. *An Inquiry into the Human Prospect*. New York: W. W. Norton.

Heimer, Carol A. 1988. "Social Structure, Psychology, and the Estimation of Risk." *Annual Review of Sociology* 14:491–519.

Hemmingway, Roy. 1983. "The Northwest Power Planning Council: Its Origins and Future Role." *Environmental Law* 13:673–97.

Hilborn, Ray. 1987. "Living with Uncertainty in Resource Management." *North American Journal of Fisheries Management* 7:1–5.

———. 1990. Personal communication to the author.

———. 1992. "Institutional Learning and Spawning Channels for Sockeye Salmon (*Oncorhynchus nerka*)." *Canadian Journal of Fisheries and Aquatic Sciences* 49:1126–36.

Hirschman, Albert O. 1970. *Exit, Voice, and Loyalty: Responses to Decline in Firms, Organizations, and States*. Cambridge, Mass.: Harvard University Press.

Hirst, Eric, Richard Goeltz, and David Trumble. 1987. *Electricity Use and Savings: Final Report, Hood River Conservation Project*. Oak Ridge, Tenn.: Oak Ridge National Laboratory, ORNL/CON-231, DOE/ BP-11287-16.

Holling, C. S., ed. 1978. *Adaptive Environmental Assessment and Management*. New York: John Wiley & Sons.

Holzner, Burkhart, and John Marx. 1979. *Knowledge Application: The Knowledge System in Society*. Boston: Allyn and Bacon.

Hughes, Thomas P. 1983. *Networks of Power: Electrification in Western Society, 1880–1930.* Baltimore: Johns Hopkins University Press.

Kahneman, Daniel, Paul Slovic, and Amos Tversky. 1982. *Judgment under Uncertainty: Heuristics and Biases.* Cambridge: Cambridge University Press.

Kahrl, William, ed. 1978. *California Water Atlas.* Los Altos: William Kauffmann.

Kennan, George F. 1972. *Memoirs.* Vol. 2. Boston: Little, Brown.

Keto, Aila I., and Keith Scott. 1986. *Logging Is Incompatible with Conservation of North Queensland's Tropical Rainforests.* Rainforest Conservation Society of Queensland.

Kingdon, John W. 1984. *Agendas, Alternatives, and Public Policies.* Boston: Little, Brown.

Koomey, Jonathan Garo. 1990. "Energy Efficiency Choices in New Office Buildings: An Investigation of Market Failures and Corrective Policies." Ph.D. diss., Energy and Resources Group, University of California, Berkeley.

Landau, Martin. 1969. "Redundancy, Rationality, and the Problem of Duplication and Overlap." *Public Administration Review* 19 (July/August):346–58. Reprinted in *Bureaucratic Power in National Politics*, ed. Francis E. Rourke. 3d ed. Boston: Little, Brown, 1978.

Landau, Martin, and Russell Stout, Jr. 1979. "To Manage Is Not to Control, or the Folly of Type II Errors." *Public Administration Review* 39:148–56.

Landy, Marc K., Marc J. Roberts, and Stephen R. Thomas. 1990. *The Environmental Protection Agency: Asking the Wrong Questions.* New York: Oxford University Press.

LaPorte, Todd R., and Paula M. Consolini. 1991. "Working in Practice but Not in Theory: Theoretical Challenges of High-Reliability Organizations." *Journal of Public Administration Research and Theory* 1:19–47.

Lawrence, Jody. 1983. "The Water Budget: A Step towards Balancing Power and Fish in the Columbia River Basin." M.S. thesis, Department of Civil Engineering, University of Washington.

Lee, Kai N. 1983. "The Path along the Ridge: Regional Planning in the Face of Uncertainty." *Washington Law Review* 58:317–42.

———. 1989. "The Mighty Columbia: Experimenting with Sustainability." *Environment* 31, no. 6 (July/August): 6–11, 30–33.

———. 1991a. "Rebuilding Confidence: Salmon, Science, and Law in the Columbia Basin." *Environmental Law* 21:745–805.

———. 1991b. "Unconventional Power: Energy Efficiency and Environmental Rehabilitation under the Northwest Power Act." *Annual Review of Energy and the Environment* 16:337–64.

Lee, Kai N., and Donna Lee Klemka, with Marion E. Marts. 1980. *Electric Power and the Future of the Pacific Northwest.* Seattle: University of Washington Press.

Lee, Kai N., and Jody Lawrence. 1986. "Adaptive Management: Learning from the Columbia River Basin Fish and Wildlife Program." *Environmental Law* 16:431–60.

Leigland, James, and Robert Lamb. 1986. *WPPSS: Who Is to Blame for the WPPSS Disaster.* Cambridge, Mass.: Ballinger.

Lempert, Richard O., and Christy A. Visher. 1988. "Randomized Field Experiments in Criminal Justice Agencies." *Research in Action* (National Institute of Justice, U.S. Department of Justice newsletter). October.

Leopold, Aldo. 1949. *A Sand County Almanac.* Reprint, New York: Ballantine, 1970.

Levinthal, Daniel and James G. March. 1981. "A Model of Adaptive Organizational Search." *Journal of Economic Behavior and Organization* 2:307–33. Reprinted in March, *Decisions and Organizations,* Oxford: Basil Blackwell, 1988, 187–218. (Page citations follow the latter version.)

Levitt, Barbara, and James G. March. 1988. "Organizational Learning." *Annual Review of Sociology* 14:319–40.

Lilienthal, David E. 1943. *TVA: Democracy on the March.* Reprint: New York: Harper & Brothers, 1953.

Lindblom, Charles E. 1959. "The Science of 'Muddling Through.' " *Public Administration Review* 19 (Spring):79–88.

Liverman, Diana M., et al. 1988. "Global Sustainability: Toward Measurement." *Environmental Management* 12:133–43.

Lothrop, Robert C. 1986. "The Misplaced Role of Cost-Benefit Analysis in Columbia Basin Fishery Mitigation." *Environmental Law* 16:517–54.

Mahar, Dulcy, and Mickey Riley. 1985. "Interview: Chuck Collins." *Northwest Energy News,* January/February, 6–9.

March, James G. 1972. "Model Bias in Social Action." *Review of Educational Research* 42:413–29. Reprinted in March and Olsen 1976.

———. 1988. *Decisions and Organizations.* Oxford: Basil Blackwell.

March, James G., and Johan P. Olsen. 1976. *Ambiguity and Choice in Organizations.* Bergen, Norway: Universitetsforlaget.

———. 1984. "The New Institutionalism: Organizational Factors in Political Life." *American Political Science Review* 78:734–49.

———. 1989. *Rediscovering Institutions: The Organizational Basis of Politics.* New York: Free Press.

Mathews, Jessica Tuchman. 1989. "Redefining Security." *Foreign Affairs* 68:162–77.

McKinley, Charles. 1952. *Uncle Sam in the Pacific Northwest.* Berkeley: University of California Press.

Miles, Edward L. 1987. *Science, Politics, and International Ocean Management.* Policy Papers in International Affairs, no. 33. Berkeley: University of California Institute of International Studies.

Moe, Terry M. 1984. "The New Economics of Organization." *American Journal of Politics* 28:739–77.

Muckleston, Keith W. 1990. "Salmon vs. Hydropower: Striking a Balance in the Pacific Northwest." *Environment*, January/February, 10–15, 32–36.

Nader, Laura, and Harry F. Todd, Jr., eds. 1978. *The Disputing Process: Law in Ten Societies*. New York: Columbia University Press.

National Research Council. 1986. *Ecological Knowledge and Environmental Problem-Solving*. Report of the Committee on the Applications of Ecological Theory to Environmental Problems. Washington, D.C.: National Academy Press.

Nectoux, François, and Yoichi Kuroda. 1990. *Timber from the South Seas: An Analysis of Japan's Tropical Timber Trade and Its Environmental Impact*. Gland, Switzerland: World Wildlife Fund International.

Northwest Power Planning Council. 1982. "To Save the Salmon." *Northwest Energy News* 1, no. 3 (May/June).

———. 1987a. *Columbia River Basin Fish and Wildlife Program*. Portland, Ore.

———. 1987b. "A Review of Conservation Costs and Benefits. Five Years of Experience under the Northwest Power Act." Issue paper. Portland, Ore.

———. 1988. "Protected Areas Amendments and Response to Comments." Issue paper. Portland, Ore.

———. 1988–1990. Minutes, System Planning Oversight Committee. Portland, Ore.

———. 1989. "Assessment of Regional Progress toward Conservation Capability Building." Issue paper. Portland, Ore.

———. 1990. "Yakima Production Project: Review of Preliminary Design Report." Issue paper. Portland, Ore.

———. 1991. *Northwest Conservation and Electric Power Plan*. Portland, Ore.

Olson, Mancur, Jr. 1965. *The Logic of Collective Action*. Cambridge, Mass.: Harvard University Press.

Orians, Gordon H. 1980. "Aggregations: Curse and Necessity." In *The National Research Council, 1980: Issues and Current Studies*. Washington, D.C.: National Academy Press.

Pacific Northwest Utilities Conference Committee, PNUCC Task Force 8. 1977. "Report to the Technical Coordination Group." Portland, Ore.

Park, David. 1988. *The How and the Why: An Essay on the Origins and Development of Physical Theory*. Princeton: Princeton University Press.

Pearce, David W. 1988. "Sustainable Use of Natural Resources in Developing Countries." In *Sustainable Environmental Management: Principles and Practice*, ed. R. Kerry Turner. Boulder: Westview Press.

Pearce, David W., Edward B. Barbier, and Anil Markandya. 1988. "Sustainable Development and Cost Benefit Analysis." LEEC Paper 88-03, International Institute for Environment and Development. London.

Perrin, Noel. 1979. *Giving Up the Gun*. Boston: David R. Godine.

Peterman, Randall M. 1990. "Statistical Power Analysis Can Improve Fisheries Research and Management." *Canadian Journal of Fisheries and Aquatic Sciences* 47:2–15.

Philips, Marion, et al. 1987. "Cost Analysis: Final Report, Hood River Conservation Project." Bonneville Power Administration, U. S. Department of Energy, DOE/BP-11287-8.

Platt, John, and Doug Dompier. 1990. "A History of State and Federal Fish Management in the Columbia and Snake River basins." *CRITFC News* (Columbia River Inter-Tribal Fish Commission). July, 9–11.

Poore, Duncan, et al. 1989. *No Timber without Trees: Sustainability in the Tropical Forest*. London: Earthscan Publications.

Pratt, John W., and Richard J. Zeckhauser. 1985. *Principals and Agents: The Structure of Business*. Boston: Harvard Business School Press.

Price, Don K. 1965. *The Scientific Estate*. New York: Oxford University Press.

Putnam, Robert D. 1988. "Diplomacy and Diplomatic Politics: The Logic of Two-Level Games." *International Organization* 47 (Summer):427–60.

Redclift, Michael. 1987. *Sustainable Development: Exploring the Contradictions*. London and New York: Methuen.

Regier, H. A., and G. L. Baskerville. 1986. "Sustainable Redevelopment of Regional Ecosystems Degraded by Exploitive Development." In Clark and Munn 1986.

Reisner, Marc P. 1986. *Cadillac Desert*. New York: Viking.

Rivlin, Alice M. 1971. *Systematic Thinking for Social Action*. Washington, D.C.: Brookings Institution.

Ruckelshaus, William. 1985. "Risk, Science, and Democracy." Speech to the National Academy of Sciences, Washington, D.C., 22 June 1983. *Issues in Science and Technology*, Spring, 19–20.

Ruderman, Henry, Mark D. Levine, and James E. McMahon. 1987. "The Behavior of the Market for Energy Efficiency in Residential Appliances Including Heating and Cooling Equipment." *Energy Journal* 8:101–24.

Sabatier, Paul A. 1988. "An Advocacy Coalition Framework of Policy Change and the Role of Policy-Oriented Learning Therein." *Policy Sciences* 21:129–68.

Sahlins, Marshall. 1972. *Stone Age Economics*. Chicago: Aldine-Atherton.

Savory, Allan. 1988. *Holistic Resource Management*. Washington, D.C.: Island Press.

Schalk, Randall. 1986. "Estimating Salmon and Steelhead Usage in the Columbia Basin before 1850: The Anthropological Perspective." *Northwest Environmental Journal* 2:1–30.

Schumpeter, Joseph A. 1942. *Capitalism, Socialism, and Democracy*. Reprint, New York: Harper Torchbooks, 1962.

Scott, Richard W. 1987. *Organizations: Rational, Natural, and Open Systems*. 2d ed. Englewood Cliffs, N.J.: Prentice-Hall.

Scoullos, Michael. 1991. Interview with the author. Osaka, 15 January.

Seattle Master Builders Association et al. v. Pacific Northwest Electric Power and Conservation Planning Council, 786 F.2d 1359 (1986, U.S. Court of Appeals, 9th Cir).

Selznick, Philip. 1957. *Leadership in Administration*. New York: Harper & Row.

Sesser, Stan. 1991. "A Reporter at Large (The Borneo Rain Forest)." *New Yorker*, 27 May, 42–67.

Shapiro, Martin. 1975. "Courts." In *Handbook of Political Science*, ed. Fred I. Greenstein and Nelson W. Polsby. Vol. 5. Reading, Mass.: Addison-Wesley.

Silk, Leonard. 1989. "Rich and Poor: The Gap Widens." *New York Times*, 12 May, C2.

Simon, Herbert A. 1954. *Administrative Behavior*. New York: Macmillan.

———. 1983. *Reason in Human Affairs*. Stanford: Stanford University Press.

Sims, C. W., and F. J. Ossiander. 1981. "Migration of Juvenile Chinook Salmon and Steelhead Trout in the Snake River Basin from 1973 through 1979: A Research Summary." Report to the U.S. Army Corps of Engineers under Contract DACW68-78-C-0038. Portland, Ore.

Sorenson, Theodore C. 1963. *Decision-Making in the White House: The Olive Branch or the Arrows*. New York: Columbia University Press.

Stegner, Wallace. 1954. *Beyond the Hundredth Meridian: John Wesley Powell and the Second Opening of the West*. Reprint, Lincoln: University of Nebraska Press, 1982.

———. 1983. "Inheritance." In Wallace Stegner and Page Stegner, *American Places*. Moscow: University of Idaho Press.

Steinbruner, John D. 1974. *The Cybernetic Theory of Decision*. Princeton: Princeton University Press.

Sugai, Wayne H. 1987. *Nuclear Power and Ratepayer Protest: The Washington Public Power Supply Crisis*. Boulder: Westview Press.

Taubes, Gary. 1992. "A Dubious Battle to Save the Kemp's Ridley Sea Turtle." *Science* 256 (1 May): 614–16.

Taylor, Serge. 1984. *Making Bureaucracies Think*. Stanford: Stanford University Press.

Tennessee Valley Authority v. Hill et al., 437 U.S. 153, 178 (1977), quoting U.S. House of Representatives Report no. 412, 93d Cong., 1st sess., 1973, 4.

Tietenberg, Thomas H. 1992. *Environmental and Natural Resource Economics*. 3d ed. New York: HarperCollins.

Thompson, James D., and Arthur Tuden. 1959. "Strategies, Structures, and Processes of Organizational Decision." In *Comparative Studies in Administration*. Pittsburgh: University of Pittsburgh Press.

Timber, Fish, and Wildlife Final Agreement. 1987. "Timber/Fish/Wildlife—A Better Future in Our Woods and Streams." Seattle.

Tollefson, Gene. 1987. *BPA and the Struggle for Power at Cost.* Washington, D.C.: U.S. Government Printing Office.

Trigger, Bruce G. 1991. "Early Native North American Responses to European Contact: Romantic versus Rationalistic Interpretations." *Journal of American History* 77:1195–1215.

Truman, David B. 1971. *The Governmental Process.* 2d ed. New York: Knopf.

Tufte, Edward R. 1974. *Data Analysis for Policy and Politics.* Englewood Cliffs, N.J.: Prentice-Hall.

Turner, R. Kerry. 1988. *Sustainable Environmental Management Principles and Practice.* Boulder, Colo.: Westview Press.

United Nations Environment Programme. 1989. Statement on Sustainable Development, adopted by Governing Council at fifteenth session, 23 May 1989. Quoted in International Tropical Timber Organization, *ITTO Guidelines for the Sustainable Management of Natural Tropical Forests,* ITTO Technical Series 5, December 1990, app. 5.

United Nations World Commission on Environment and Development. 1987. *Our Common Future.* New York: Oxford University Press.

U.S. Congress. 1980. Public Law 96-501, 16 U.S.C. 839a et seq., 94 Stat. 2697-2736. Pacific Northwest Electric Power Planning and Conservation Act.

U.S. Department of Energy, Bonneville Power Administration. 1987. *The Hood River Story.* Vol 1: *How a Conservation Project Was Implemented,* ed. Karen Schoch. DOE/BP-11287-12. Vol. 2: *Marketing a Conservation Project,* ed. Shellie Kaplon. DOE/BP-11287-13. Portland, Ore.

————. 1989a. *Forecast of Electricity Use in the Pacific Northwest.* DOE/BP-1262.

————. 1989b. *1989 Financial Summary.* DOE/BP-1318.

————. 1990. *Big Savings from Small Sources: How Conservation Measures Up.* DOE/BP-1292. Portland, Ore.

U.S. General Accounting Office. 1987. *Federal Electric Power: A Five-Year Status Report on the Pacific Northwest Power Act.* Washington, D.C. RCED-87-6.

Volkman, John M. 1992. "Making Room in the Ark: The Endangered Species Act and the Columbia River Basin." *Environment* 34, no. 4 (May):18–20, 37–43.

Volkman, John M., and Kai N. Lee. 1988. "Within the Hundredth Meridian: Western States and Their River Basins in a Time of Transition." *University of Colorado Law Review* 59:551–77.

Walters, Carl. 1986. *Adaptive Management of Renewable Resources.* New York: Macmillan.

Weinberg, Alvin M. 1972. "Science and Trans-Science." *Minerva* 10:209.

Whitehead, Alfred North. 1925. *Science and the Modern World.* 1948 edition cited in Gifford, *The Farther Shore* (New York: Vintage, 1991), p. 50.

Wildavsky, Aaron. 1979. *Speaking Truth to Power.* Boston: Little, Brown.

Wilkinson, Charles F. 1989. "Aldo Leopold and Western Water Law: Thinking Perpendicular to the Prior Appropriation Doctrine." *Land and Water Law Review* 24:1–38.

Wilkinson, Charles F., and D. Keith Conner. 1983. "The Law of the Pacific Salmon Fishery." *University of Kansas Law Review* 32:17–109.

Wolf, Charles P. 1979. "A Theory of Nonmarket Failure: Framework for Implementation Analysis." *Journal of Law and Economics* 22:107–39.

Woodhouse, Edward J. 1988. "Sophisticated Trial and Error in Decision Making about Risk." In *Technology and Politics*, ed. Michael E. Kraft and Norman J. Vig. Durham: Duke University Press.

World Bank. 1992. *World Development Report, 1992: Development and the Environment*. New York: Oxford University Press.

Worldwatch Institute. 1987. *State of the World: A Worldwatch Institute Report on Progress toward a Sustainable Society*. New York: W. W. Norton.

Worster, Donald. 1985. *Rivers of Empire: Water, Aridity, and the Growth of the American West*. New York: Pantheon.

_____. 1990a. "Transformations of the Earth: Toward an Agroecological Perspective in History." *Journal of American History* 76:1087–1106.

_____. 1990b. "Seeing beyond Culture." *Journal of American History* 76:1142–47.

Acknowledgments

When Jan Carpenter showed me a review she had written of *Our Common Future*, a 1987 study by the United Nations World Commission on Environment and Development, I grasped for the first time the connection between what we were trying to do in the Columbia basin and the paradoxical concept of sustainable development. In years since then, she and our friend and colleague John M. Volkman of the Northwest Power Planning Council—whose writing on sustainable development influenced my own—have criticized innumerable drafts, talked me out of silly ideas, and kept our shared interest in the subjects of this book alive and lively. I also acknowledge the contributions that Todd R. LaPorte, Ernst B. Haas, Richard Rabinowitz, W. Lance Bennett, and Steve W. Berman have made to the way I think.

Four colleagues at the University of Washington supplied essential challenges: Gordon H. Orians, a distinguished ecologist, suggested that I write a book on how adaptive management could be turned into practical public policy; Edward L. Miles, scholar of international relations and organizational guerrilla of extraordinary *élan*, doubted that anything so fragile as adaptive management could survive the rough-and-tumble of politics as he knew it; Gardner M. Brown, Jr., an astutely rigorous economist, challenged my too-facile ideas about sustainable economies and intertemporal comparisons, and changed the way I think about economics; and Ray Hilborn, one of the creators of the scientific ideas and modeling methods that constitute adaptive management, helped by example and empirical effort to hold off our colleagues' skepticism. This book is an answer to each of these friends, to whom I remain indebted.

At various stages, Langdon Winner, Stephen J. DeCanio, Tom Tietenberg, Peter J. May, Carl Walters, Randall M. Peterman, William C.

Clark, Kirk Jeffrey, Gerald W. Cormick, Don Gifford, Vim Wright, Toshiko Akiyama, Stephen O. Andersen, Anthony M. Starfield, Edward J. Woodhouse, Harlan Wilson, Kimberly A. Jordan, Wendy Penner, Arild Underdal, and Peter M. Haas supplied valuable advice and assistance. Mara Bún provided indispensable background on the Australian rain forest. C. S. Holling, who leads the global effort to learn from ecological experience, has been both helpfully critical and grandly inspiring. To Buzz Holling, Garry D. Brewer, and Charles F. Wilkinson, who have given sensible counsel and unstinting support, I owe the sense of optimism that underlies the arguments offered here. The Northwest Power Planning Council staff and its executive director, Edward Sheets, have contributed in ways too numerous to list to the work described in Chapter 2.

An untidy secret of education is that teachers usually learn more from students than the other way round. John N. Winton, Amanda Azous, William L. Matthaei, the late Jody Lawrence, Dori C. Cahn, Greg Hill, Wayne H. Sugai, Douglas Amy, Natalie Stevens, Hans Olav Ibrekk, and Cindy L. Halbert have each benefited me and this book. At Williams College I have had extraordinary help from Sandra L. Zepka, Marcella Rauscher, and Nan Jenks-Jay. Colleagues and students at Williams have revived my belief that civility, rigor, and integrity still grace higher education; here learning lies at the center of work.

Not being a gentleman scholar, I have incurred multiple debts of a material sort. The Washington Sea Grant Program, a wonderfully hardy exemplar of improving nationally by investing locally, supported a project that Ray Hilborn and I directed, including a substantial portion of the work described in this book. I drafted the manuscript during a sabbatical year at the Kyoto Institute of Economic Research, Kyoto University, while Maria Coronado Hernandez provided superior research assistance back in the United States. To Professor Tsuneo Tsukatani, the Ministry of Education of Japan, and Masako Ueda, I offer heartfelt thanks for generous teaching in a year full of learning.

The Banff Centre, International Institute for Applied Systems Analysis, Yosemite National Institutes, World Resources Institute, Kyoto University's Institute of Economic Research, South Florida Water Management District, University of Florida, National Geographic Society, Darden Graduate School of Business at the University of Virginia, and the Society of Alumni of Williams College offered friendly settings in which to try out many of my ideas. Earlier versions of parts of this book have appeared in the *Annual Review of Energy and Environment, Environmental Law,* and the *Northwest Environmental Journal.* I am grateful to these journals for permission to reprint this material. The Northwest Power Planning Council staff provided most of the illustrations in this book.

A book, once drafted, must seek its fortune in an overcrowded, apparently hostile world. Howard Boyer, then at Harvard University Press, became its first champion and advocate. Robert O. Keohane, whose reach and depth as a scholar of international relations are justly famed, read my draft with care and recommended restructuring the argument along the lines it now takes. Later, Joseph Ingram of Island Press put me in the rigorous and discerning hands of Ann Hawthorne and Barbara Youngblood.

The deepest debts cannot be counted or accounted, because they are constitutive: in accumulating them, I have become who I am. This book is for Katherine and Anna Lee, for Hsin Chih Lee and the memory of Chofeng Lin Lee, for Kirk and Virginia Jeffrey—and, once more, for Dana.

Index

About the Author

Kai N. Lee is professor of environmental studies and
director of the Center for Environmental Studies at
Williams College in Williamstown, Massachusetts. Lee
was educated in physics at Columbia (A.B., 1966) and
Princeton (Ph.D., 1971) and did postdoctoral work in
political science at the University of California, Berkeley.
From 1973 until 1991 he taught environmental studies and
political science at the University of Washington in Seattle.
In 1976–77 he served as a White House Fellow at the U.S.
Department of Defense and from 1983 until 1987 Lee
represented the State of Washington on the Northwest
Power Planning Council. Lee has served on several
committees of the National Research Council and is now
a member of its Board on Environmental Studies and
Toxicology.

broadview press

Post Office Box 1243 · Peterborough · Ontario · K9J 7H5
Telephone: (705)743-8990 · Fax: (705)743-8353

Please be informed that
our current price on
Compass & Gyroscope is $34.95

Thank you.
for your interest
B.
comps dept.